橋·樑·工·程

徐耀賜 博士 著

全華科技圖書股份有限公司 印行

落實西方科技

本土化

之第一步工作

乃是專業名詞

中文化之

統一。

~徐耀賜~

自序

筆者自1982年秋負笈美國馬里蘭大學土木工程研究所以來，即因研究與工作之便而收集了無數與橋樑有關之名詞。今將這些名詞以最淺顯之中英對照方式整理問世，冀望能對橋樑工程專業名詞中文化與本土化有微薄之貢獻。

本人原先之構想除了中英對照之外，每一個名詞亦想輔以深入之解釋與圖片解說，並據此向相關單位申請研究計畫，惟筆者之熱勁卻無法得到迴響。是以，吾人只能以最淺顯簡易之方式作此編排，雖不具所謂之「學術價值」，惟吾人相信對科技本土化必具有「拋磚引玉」之功用。

與橋樑相關之工程名詞甚廣，本書收錄者必有掛失，尚祈先進不吝指正。

晚
徐耀賜謹誌

編輯部序

「系統編輯」是我們的編輯方針，我們所提供給您的，絕不只是一本書，而是關於這門學問的所有知識，它們由淺入深，循序漸進。

本書內容將橋樑相關之工程名詞依 A、B、C、……Y 字母順序排列，並以中英之對照方式，使讀者易於從第一字母查詢所需之單字。凡是對橋樑工程有興趣者，皆可利用此書提昇自己專業術語的能力，使您在工作及學業上，更能發揮得淋漓盡致。

目 錄

英　　文	中　　文
Abrasion Resistance	混凝土表面之磨耗拉力
Absorption Theory	吸附理論
Abutment	橋台
Acceleration	加速度
Acceleration Coefficient	加速係數(橋樑耐震設計用之)
Acceleration Lane	加速度車道
Accelerator	早強劑
Acid Etching	酸蝕
Acoustic Emission Method	聲射法(非破壞性檢測之一種)
Action	作用力(Reaction之反義字)
Active Control System	主動控制系統(橋樑隔減震常用之名詞)
Active Crack	活性裂縫(與Live Crack同義)
Active Earth Pressure	主動土壓力
Active Stress Failure Plane	主應力破壞面
Active Wedge	主動土楔
Activity of Clay	黏土之活性度
Acute Corner	銳角
Admixture	摻料，混合料
Adhesive Force	黏著力

見下頁續…

英　　文	中　　文
Advancing Shoring Method (ASM)	支撐先進工法(後拉預力混凝土橋樑施工法之一種)
Age of Concrete	混凝土齡期
Aging Losses	時變損失(例如鋼筋或其他構件，因年代日增，其有效斷面愈小)
Aging Strain	時變應變
Aggradation	漲床(指河床面因有其他雜物而上升之現象，與Degradation反義)
Aggregate	骨材(例如砂、礫石等等)
Alignment	線形(指公路之幾何形狀)
Alkali-Silica Reaction	鹼骨材反應(對混凝土而言)
All-lightweight Concrete	全輕質混凝土(指輕質混凝土中不含天然砂者)
Allowable Bearing Capacity	容許承載力
Allowable Slip Resistance	容許滑動阻力
Allowable Stress	容許應力

見下頁續…

英　　文	中　　文
Allowable Stress Design (ASD)	容許應力設計(與Working Stress Design與Service Load Design同義)
Allowable Tolerance	容許尺寸誤差
Allowable Unit Stress	容許單位應力
Allowable Working Stress	容許工作應力
Alloy	合金
Alluvium Deposit	沖積層
Alternate Design Method	替代設計方法
Alternate Load Paths	替代荷載路徑
Alternative	替代方案
Aluminum Bridge	鋁橋
Aluminum Conduits	鋁導管
Aluminum Railing	鋁欄杆
Ambient Vibration	微振
American Association of State Highway and Transportation Officials (AASHTO)	美國州際公路與運輸官員協會(其前身為AASHO，1921～1973年，1973年之後更名為AASHTO)
American Concrete Institute (ACI)	美國混凝土學會
American Galvanizers Association (AGA)	美國鍍鋅業者協會

見下頁續…

英　　文	中　　文
American Institute of Steel Construction (AISC)	美國鋼結構營建協會
American Iron and Steel Institute (AISI)	美國鋼鐵學會
American Railway Engineering Association (AREA)	美國鐵路工程學會
American Segmental Bridge Institute (ASBI)	美國節塊式橋樑學會
American Society for Testing and Materials (ASTM)	美國材料與試驗協會
American Society of Civil Engineers (ASCE)	美國土木工程師學會
American Standards Association (ASA)	美國標準協會
American Traffic Safety Services Association (ATSSA)	美國交通安全服務協會
American Welding Society (AWS)	美國焊接協會
Amplification Factor	放大因數
Anchor (Anchorage)	錨碇

見下頁續…

英　　文	中　　文
Anchor Bar	錨碇棒(通常指位於混凝土Corbel處之鋼筋或鋼棒)
Anchor Bolt	錨碇螺栓
Anchor Failure	錨碇失敗，錨碇破壞
Anchor Head	錨頭(地錨系統錨碇之頭部位置)
Anchor Plate	錨碇鈑
Anchorage Access	錨碇凹槽
Anchorage Block	錨碇塊(通常用於後拉預力混凝土橋樑)
Anchorage Zone	錨碇區
Anchored Retaining Wall	錨碇式擋土牆
Anchored Walls	錨碇牆(指擋土牆具有地錨等錨碇設施)
Angle of Friction	摩擦角
Angle of Load Eccentricity	載重偏心角
Angle of Internal Friction	內摩擦角(針對土壤而言)
Angles	角度、角鋼
Angstrom	埃(射線波長量測之單位)
Anisotropy	異向性
Annual Average Daily Traffic (AADT)	年平均每日交通量

見下頁續…

英　　文	中　　文
Annual Runoff	年逕流量(橋樑排水設計用之)
Anode	陽極
Anti-Corrosion Paint	防銹漆
Anti-Corrosion Strategy	腐蝕防治策略
Anti-Friction Bearing	反摩擦支承
Applied Technology Council (ATC)	美國之應用科技組織(此單位出版甚多具有非常實用價值之應用規範，例如ATC6對耐震設計敘述甚詳)
Approach	引道(與Approach Roadway同義)
Approach Guardrail	引道護欄
Approach Roadway	引道
Approach Slab	進橋版(引道版)
Approach Slab Seat	進橋版座(通常位於橋台之背牆)
Apron	沖刷保護工(位於橋台與橋墩處防止橋樑下部結構被沖刷或淘空)
Arc Length	弧長
Arc Weld	電弧焊接
Arch	拱
Arch Bridge	拱橋

見下頁續…

英　　文	中　　文
Arch Rip	拱肋
Arc Rise	拱頂，拱矢
Architectural Institute of Japan (AIJ)	日本建築學會
Artifical Intelligence (AI)	人工智慧
Armor	加強構件，加強桿
As-Built Drawing	竣工圖
As-Built Plan	竣工圖
As-Built Shop Drawing	竣工圖
Asphalt	瀝青
Asphalt Concrete (AC)	瀝清混凝土
Asphalt Overlay	瀝青磨耗層
At-Grade Crossing	平面交叉(公路交叉型式之一種)
Atmospheric Corrosion Resisting Steel	耐候性鋼材(與Weathering Steel同義)
At-rest Earth Pressure Coefficient	靜止土壓力係數
Availability	可用性
Autostress Design	自動應力設計(AASHTO規範中鋼橋設計方法之一種)
Average Daily Traffic (ADT)	平均每日交通量
Average Daily Truck Traffic (ADTT)	平均每日卡車交通量

見下頁續…

英　文	中　文
Axial Capacity	軸向承載力
Axial Force	軸向力(軸向意指構件長度方向)
Axial Load	軸向力，軸向荷重，軸向載重
Axial Stress	軸向應力
Axle Load	軸重(針對車輛活載重而言)

英　　　文	中　　　文
Back Filler	回填料
Backfill Soil	回填土
Backwall	後牆(橋台之一部份，其後為背填土)
Backwater	回水(由於河床面有障礙物存在，使自由水水位上升高於正常水位之現象)
Bagged Concrete	混凝土包，混凝土袋
Bagged Stone	石包，石袋
Bailey Bridge	倍力橋(用於搶修工作之臨時用橋樑，軍中由工兵負責)
Balanced Cantilever Construction	平衡懸臂施工法(簡稱懸臂工法)
Balanced Cantilever Erection	平衡懸臂架設
Balanced Reinforcement Ratio	平衡鋼筋比
Balanced Strain Conditions	平衡應變狀況
Ballast	道碴
Bank Protection	土坡保護，護坡
Bar Chair	鋼筋支座
Bar Coupler	鋼棒續接器

見下頁續…

英　　　文	中　　　文
Bar Lists	鋼筋列舉(在設計圖上把鋼筋之使用量與種類列出，以方便包商估價與施工管理)
Bar Spacing	鋼腱間距，鋼棒間距
Bar Support	鋼筋支座(與Bar Chair同義)
Barge	浮動船塢
Barrier	護欄
Barrier Railing	護欄杆
Bascule Bridge	上下旋轉式橋樑(可移動式橋樑之一種)
Base Failure	基底破壞
Base Inclination Factors	基底傾斜因數
Base Isolation	基底隔震
Base Plate	底承版
Battered Piles	斜樁
Beam	樑
Beam Bridges	樑式橋
Beam-Cable Element	樑索元素
Beam-Column Structure	樑－柱結構
Beam End	樑端
Beam on Elastic Foundation (BEF)	彈性基礎樑

見下頁續…

英　　　文	中　　　文
Bearing	支承(簡稱Brg.)
Bearing Area	承載面積
Bearing Capacity	承載力
Bearing Capacity Factor	承載力因數
Bearing Pile	支承樁，承載樁
Bearing Plate	承墊版，承壓版
Bearing Reaction	支承反力
Bearing Pressure	承載壓力
Bearing Seat	支承座
Bearing Stiffeners	支承加勁鈑(位於支承處，腹版兩側)
Bearing Stress	承載應力
Bearing Surface	承載面，支承面
Bed Rock	岩床，基岩
Bedding Factor	安床因數(地下埋管設計時候用)
Bench Mark (BM)	水準點(公路測量定線用之)
Bending	彎矩，彎曲
Bending Action	彎矩作用
Bending Capacity Reduction Factor	彎矩強度折減因數
Bending Distortion	彎矩畸變
Bending Member	受彎構件

見下頁續…

英　　文	中　　文
Bending Moment	彎矩(簡稱Bending或Moment)
Bending Moment Coefficient	彎矩係數
Bending Normal Stresses	彎矩正向應力
Bending Rigidity	彎矩剛度
Bending Shear Stresses	彎矩剪應力
Bending Stresses	彎矩應力
Bending Stress Coefficient	彎矩應力係數
Bent	多柱式下部結構(由雙柱或以上形成之橋樑下部結構，柱上通常有帽樑)
Bent Caps	墩帽
Bent Piers	多柱式橋墩
Bent Plates	彎鈑(用於鋼結構中)
Biaxial Loading	雙軸載重
Biaxial Symmetrical Section	雙軸對稱斷面(例如AISC之WF斷面)
Bidding Stage	競標階段
Bimoment	雙彎矩(與翹曲彎矩，Warping Moment，同義)
Bimoment Normal Stress	雙彎矩正向應力
Bituminous	瀝青
Bituminous Concrete	瀝青混凝土

見下頁續…

英　　文	中　　文
Bituminous Overlay	瀝青磨耗層
Bleeding (Water)	浮水，泌水(濕混凝土表面之懸浮水)
Blister	凸出錨塊(用於後拉預力混凝土結構)
Bogies	台車(橋樑施工時之臨時性運輸車輛，與Carriage同義)
Bolt	螺栓
Bolted Connection	栓接(螺栓接頭，螺栓接合)
Bonding Agent	黏結劑，接著劑
Bonding Compound	黏結劑，接著劑
Bonding Stress	握裹應力(針對鋼筋而言)
Bored Piles	鑽灌樁
Boring	鑽掘
Boring Log	土壤鑽探柱狀圖
Borrow Pit	借土坑，取土坑
Bottom Chord	下弦桿
Bottom Flange	底版，下翼版
Bottom Slab	底版
Bottom Slab Reinforcement	底版鋼筋
Boundary Element Method	邊界元素法
Box Caisson	箱型沈箱

見下頁續…

英　　文	中　　文
Box Culvert	箱涵
Box Girder	箱型樑
Box Girder Inspection Walkway	箱型樑檢測步道(大型箱型樑之斷面甚大，甚至足以使人步行其間，以達到檢測之目的)
Box Pier	箱型橋墩
Box Section	箱型斷面
Box-Type Abutment	箱型橋台
Braced Non-Compact Section	具側向支撐之非堅實斷面(AASHTO之LFD鋼橋設計中之某一種斷面型式)
Bracing Members	支撐構件
Bracket	托樑，托架(大陸亦稱為牛腿)
Braided River	髮辮型河川
Braking Force	煞車力(或縱向力)
Break-Off Test Method	表面彎裂試驗法(非破壞性檢測之一種)
Breastwall	胸牆(橋台、橋墩之主體)
Brick Bridge	磚橋
Bridge Abutment	橋台(簡稱Abutment)
Bridge Approach Roadway	橋樑引道
Bridge Authority	橋樑主管機關

見下頁續…

英　　文	中　　文
Bridge Classification System (BCS)	橋樑分類系統
Bridge Clearance Diagram	橋樑淨距圖
Bridge Construction Manual	橋樑施工手冊
Bridge Deck	橋面版(簡稱Deck)
Bridge Deficiency	橋樑功能失調
Bridge Design Code	橋樑設計規範
Bridge Design Expert System (BDES)	橋樑設計專家系統
Bridge Design Firm	橋樑設計公司
Bridge Design Specifications	橋樑設計規範
Bridge Drainage System	橋樑排水系統
Bridge Electrical System	橋樑電氣系統
Bridge Engineer	橋樑工程師
Bridge Engineering	橋樑工程(日本人稱橋樑工學)
Bridge Floor	橋床版，橋面
Bridge Floor Surfaces	橋版面
Bridge ID	橋樑基本資料(ID乃是Identification之簡稱)
Bridge Inspection	橋樑檢測
Bridge Inspection Report	橋樑檢測報告

見下頁續…

英　　　文	中　　　文
Bridge Inspector	橋樑檢測員
Bridge Inventory	橋樑檔案
Bridge Layout	橋樑佈置，橋樑放樣
Bridge Length	橋長
Bridge Lighting	橋樑照明
Bridge Loading	橋樑荷重
Bridge Locations	橋樑區位
Bridge Location Plan	橋樑位置圖
Bridge Management System (BMS)	橋樑管理系統
Bridge Monitoring System	橋樑監測系統
Bridge Numbering System (BNS)	橋樑編號系統
Bridge Pad (Bridge Seat Bearing Area)	橋座支承區
Bridge Plans	橋樑圖樣(指設計圖、施工圖或竣工圖)
Bridge Rail (Bridge Railing)	橋樑欄杆
Bridge Ranking System (BRS)	橋樑排序系統
Bridge Rating	橋樑評定
Bridge Rating System (BRS)	橋樑評定系統
Bridge Rehab. Job	橋樑修復工作

見下頁續…

英　　　文	中　　　文
Bridge Relocation	橋樑改位
Bridge Seat	橋座
Bridge Site	橋址
Bridge Site Data	橋址數據
Bridge Welding Code	橋樑焊接規範
Bridge Widening	橋樑拓寬
British Iorn & Research Association (BIRA)	英國鋼鐵與研究協會
British Standards (BS)	英國國家標準(BS規範在橋樑工程領域具有甚高之參考價值)
British Standards Institution	英國標準學會
Brittle Failure	脆性破壞
Brittle Fracture Crack	脆性裂縫(指鋼結構而言)
Brittleness	脆性
Buckling	挫屈
Buckling Coefficient	挫屈係數
Buckling in Plane	面內挫屈
Buckling Stress	挫屈應力
Buffer	緩衝器(隔減震設計之用語)
Buffeting	抖振(用於橋樑風力分析)
Built-Up Member	組合構件

見下頁續…

英　　文	中　　文
Bulkhead	駁岸，隔版牆(與 Partition Wall同義)
Bundled Bars	束筋(通常指三根或以上)
Buoyance	浮力
Burlap	粗麻布
Buttress Abutment	倚壁式橋台

英　　文	中　　文
Cable	索，纜索
Cable Element	纜索元素(用於斜張橋與懸索橋之分析)
Cable-Stayed Bridge	斜張橋，斜拉橋
Cable Suspension Bridge	懸索橋，吊橋
Caisson	沈箱
Calcium Chloride	氯化鈣
Calender Day	日曆天
California Department of Transportation (Caltrans)	美國加州運輸部
Caliper	測徑規(量測構件直徑用之)
Camber	預拱，拱度
Camber Curve	預拱曲線，拱度曲線
Camber Diagram	預拱圖，拱度圖
Canal Bridge	運河橋(跨越運河之橋樑稱之)
Cantilever Abutment	懸臂式橋台
Cantilever Arm	懸臂
Cantilever Beam	懸臂樑
Cantilever Flange	懸翼版
Cantilever Girder	懸臂樑
Cantilever Retaining Wall	懸臂式擋土牆

見下頁續…

英　　　文	中　　　文
Cantilever Slabs	懸臂版
Cantilever Span	懸跨
Cantilever Tendon	懸臂鋼腱(用於連續式後拉預力橋樑)
Cantilevered Girder	懸臂樑
Cap	帽樑
Cap Beam	帽樑(簡稱Cap)
Capillary Tension	毛細張力
Capstone	帽石
Carbon Fiber Reinforce Plastic (CFRP)	塑炭鋼
Carbon Steel	碳鋼
Carbonation	碳酸化(針對混凝土結構而言，亦可稱中性化)
Cast-in-Place (CIP)	場鑄(即現場澆注)
Cast-in-Place Concrete Pile	場鑄混凝土樁
Cast-in-Place Joint	場鑄接頭，場鑄接縫
Cast-in-Place Post-Tensioned Bridges	場鑄後拉預力橋樑
Cast-in-Place Segment	場鑄節塊
Cast-in-Place Splice	場鑄接頭(針對預鑄混凝土橋樑而言)
Cast Iron	鑄鐵
Cast Steel	鑄鋼

見下頁續…

英　　　　文	中　　　　文
Casting Cell	澆注室
Casting Curve	澆鑄曲線
Casting Sequence	混凝土澆灌順序
Casting Yard	預鑄場
Catch Basin	截水槽
Catch Drain	排水溝
Catenary	懸索自重曲線
Cathode	陰極
Cathode Protection	陰極保護
Cavitation	孔蝕
Cellular Abutment	地窖式橋台(橋台主壁體、翼牆與引道版之間無填土存在，類似地窖，因之得名。與Vaulted Abutment同義)
Cement	水泥
Cement Mortar	水泥砂漿(水泥、4倍之砂與水形成之混合物
Cement Paste	水泥漿(水泥與水混合後之塑性漿)
Center Span	中央跨距
Centerline	中心線
Centerline Bearing	支承中心線
Centrifugal Force	離心力
Centroid	形心

見下頁續…

英　　　文	中　　　文
Centroidal Axis	形心軸
Certified Bridge Inspector (CBI)	具有證照之橋樑檢測師
Chain Drag	鐵鍊拖拉(用於偵測混凝土版不密實之部份)
Change Order	變更設計
Channel	渠道
Channel Contraction Scour Depth	河道束縮沖刷深度
Channel Profile	渠道縱坡面
Charpy Impact Test	衝擊試驗(針對金屬之韌性而言)
Check Hammer	檢查鎚，檢測鎚(橋樑檢測敲擊用之)
Check List	檢查條例
Cheekwall	邊牆(位於橋台最外側大樑之兩側，其主要之目的在於擋風遮雨，與Knee Wall同義)
Chief Engineer	總工程師，總工程司
Chinese National Standardshloride	中國國家標準(簡稱CNS)
Analysis	氯分析
Chloride Attack	鹽害
Chloride Contamination	氯化物污染或鹽污染

見下頁續…

英　　文	中　　文
Chloride Content Analysis	氯含量分析(對混凝土結構而言)
Chlorinated Rubber	氯化橡膠
Chord	弦
CIP Joint	現場接縫(CIP為Cast-in-Place之簡稱)
Circular Arch	圓拱
Circular Footing	圓形基腳
Civil Engineer	土木工程師
Civil Engineering	土木工程
Classes of Loading	載重等級
Classification of Soil	土壤分類
Clay	黏土
Clear Distance	淨距
Clearance	淨距
Clear Headway	橋下河道淨高
Clear Height	淨高
Clear Span	淨跨距
Clip Corner, Clip Angle	鋼鈑之切角，截角
Close Caisson	封口沈箱
Closed Abutment	封閉式橋台
Closed Section	封閉式斷面
Closing (Closure)	閉合
Closure Joint	閉合接頭

見下頁續…

英　　文	中　　文
Closure Segment	閉合節塊
Coarse Aggregate	粗骨材
Coating	表面塗裝，塗裝
Coating Inspection	塗裝檢測
Codes	規範
Coefficient for Pier Nose Inclination	橋墩鼻頭傾斜係數(指臨上游處)
Coefficient of Bearing Capacity	承載力係數
Coefficient of Compressibility	壓縮性係數，壓縮係數
Coefficient of Earth Pressure	土壓力係數
Coefficient of Friction	摩擦係數
Coefficient of Ground Reaction	地盤反力係數
Coefficient of Kinematic Viscosity	動黏滯係數
Coefficient of Secondary Compression	次壓力係數
Coefficient of Side Friction	側摩擦係數
Coefficient of Thermal Expansion	熱膨脹係數
Cofferdam	圍堰

見下頁續…

英　　文	中　　文
Coffered T-Beam	合成T型樑
Cohesive Soil	黏性土壤
Cold Bending	冷彎法(將熱軋型鋼彎曲之方法之一)
Cold Formed Steel	冷軋鋼(例如一般之汽車鈑金即屬此類)
Cold Joint	冷縫
Cold Mechanical Strengthtening	冷作直化(鋼結構維修方法之一)
Collision Damage	撞擊損壞(橋樑常見破壞現象之一)
Collision Load	撞擊力
Column	柱
Cohesionless Soil	非黏滯性土壤
Cohesive Soil	黏性土壤
Combination of Loads	載重組合
Combined Pile	組合樁
Combined Response Coefficient	組合反應係數
Combined Stresses	組合應力
Combined Testing Method	非破壞檢測中之綜合檢測法
Combined Footing	組合式基腳
Compact Section	堅實斷面，結實斷面
Compaction	夯實

見下頁續…

英　　文	中　　文
Composite Action	複合作用，合成作用
Composite Concrete Flexural Members	複合式混凝土抗彎構件
Composite Construction	複合式施工
Composite Damping Ratio	複合阻尼比(用於隔震橋樑之分析)
Composite Girders	複合樑，合成樑
Composite Hybrid Girders	複合式混合樑
Composite Section Modulus	複合斷面模數
Composite Steel Bridges	複合式鋼橋，合成式鋼橋
Compression Buckling	壓力挫屈
Compression Flanges	受壓翼版
Compression Index	受壓指數
Compression Members	受壓桿件，受壓構件
Compression Ratio	壓縮比
Compression Rebar	壓力鋼筋
Compression Reinforcement	壓力鋼筋
Compression Seal Joint	壓縮式填縫
Compression Strength	受壓強度，耐壓強度
Compression Zone	壓力區
Compressive Strength Test	壓力強度試驗

見下頁續…

英　　文	中　　文
Compressive Stress Path	壓應力路徑
Computer Aided Design (CAD)	電腦輔助設計
Computer Aided Design and Drafting (CADD)	電腦輔助設計與繪圖
Concentrated Load	集中載重，集中荷重
Concrete Barrier	混凝土護欄
Concrete Bearing Pad	混凝土支承墊版(其上方為支承主體)
Concrete Bonding Agent	混凝土接著劑
Concrete Box Girder	混凝土箱型樑
Concrete Breaker	混凝土碎裂機
Concrete Bridge	混凝土橋
Concrete Carbonation Test	混凝土碳酸化(或中性化)試驗
Concrete Core Sample	混凝土鑽心試樣
Concrete Core Testing	混凝土鑽心試驗
Concrete End Diaphragm	混凝土端隔樑(隔版)
Concrete Hinge	混凝土支承
Concrete Mix Design	混凝土混合設計
Concrete Piles	混凝土樁
Concrete Pile Bent	混凝土樁排(其功用與橋墩相同)
Concrete Placing Sequence	混凝土澆注順序

見下頁續…

英　　　文	中　　　文
Concrete Reinforing Steel Institute (CRSI)	美國之鋼筋學會
Concrete Slab	混凝土版
Concrete Substructure	混凝土下部結構
Concrete Superstructure	混凝土上部結構
Concrete Vibrator	混凝土震動器
Condition Rating	狀況評定(橋樑評定方式之一)
Conduit	導管(橋上通常作為輸電之用)
Cone Penetration Resistance	圓錐貫入阻力
Cone Penetrometer Tests	圓錐貫入試驗(工程界以CPT簡稱之)
Connection	接頭(不同結構元件之連接部份)
Connection Plate	連接鈑(用於鋼構與鋼構之相接部份)
Consolidation	壓密(亦稱固結)
Consolidation Pressure	壓密壓力
Consolidation Settlement	壓密沈陷
Construction Equipment	施工機具
Construction Joint	構造接縫，施工縫，工作縫
Construction Load	施工荷重，施工載重

見下頁續…

英　　　　文	中　　　　文
Construction Live Load	施工活荷重
Construction Management	施工管理，營建管理
Construction Method	施工方法
Construction Overload	施工超載
Construction Sequence	施工順序
Construction Staging	分段施工
Consultant	顧問
Consulting Company	顧問公司
Contact Pressure	接觸壓力
Contact Stress	接觸應力(指基礎與土壤間之接觸應力，與Allowable Uniform Bearing Pressure同義)
Contaminant	污染物(通常指會造成橋樑結構體腐蝕之污物)
Continuing Load	持續性荷重
Continuity Tendon	連續式鋼腱
Continuous Bridge	連續式橋樑
Continuous Footing	連續式基腳
Continuous Girder	連續樑
Continuous Heating	連續式加熱法(利用加熱使軋型鋼樑彎曲之方法之一)
Continuous Weld	連續焊接(沿著接頭之全長均有焊接者)

見下頁續…

英　　　文	中　　　文
Contraction Joint	縮縫
Contractor	承包商，包商
Contriction Scour	束縮沖刷(亦有人以Contraction Scour稱之)
Controlled Permeability Formliner (CPF)	可控制滲透型模版
Cope, Coping	截角，鋼鈑於尖銳部份之切角
Copper Alloy	銅合金
Core	鑽心試樣
Core Dimension	核心尺寸
Core Drill	空心鑽(取圓柱試體時用之)
Core Form	內模
Coring Machine	混凝土鑽心機(鑽心取樣用)
Corner Crack	角落裂縫
Correction of the River Channel	為使過水橋樑處在最有利狀況下所進行之河道改變工事(例如截彎取直)
Corrosion	腐蝕，銹蝕
Corrosion Index	腐蝕指數
Corrosion Inhibiting Grease	防銹油脂
	防蝕層

見下頁續…

英　　　文	中　　　文
Corrosion Inhibitor	防蝕層
Corrosion Mechanism	腐蝕機構
Corrosion Protection System	腐蝕保護系統
Corrosion Rate	腐蝕率
Corrosion Sensor	腐蝕感應計
Corrugated Aluminum Pipe	皺形鋁管
Corrugated Duct	皺形套管(預力混凝土套管之一種)
Corrugated Steel	皺形鋼
Corrugated Steel Pipe Culvert	皺形鋼管涵
Corrugation	皺褶(通常指鋼鈑表面)
Cosmetic Treatment	表面處理
Counter Sunk Rivet	反鎖式鉚釘
Counterfort Abutment	扶壁式橋台
Counterfort Retaining Wall	扶壁式擋土牆
Couplant	耦合劑
Coupler	續接器(用於鋼筋－鋼筋與鋼腱－鋼腱間之連接
Couplings	續接器
Cover	保護層，覆蓋層

見下頁續…

英　　　　文	中　　　　文
Cover Meter	鋼筋之混凝土保護層量測器
Cover Plates	蓋鈑(用於鋼結構)
Crack	裂縫，龜裂，裂紋
Crack Arrester Hole	止裂孔(在現存鋼結構裂縫之盡頭處鑽一小孔，以防止裂縫之擴展)
Crack Bridging Capacity (CBC)	裂縫封閉能力
Crack Control	裂縫控制
Crack Control Coefficient	裂縫控制係數
Crack Depth	裂縫深度
Crack Gauge	裂縫計
Crack Inhibiting Layout	裂縫防止佈置
Crack Initiation	裂縫發生初始階段
Crack Injection	裂縫注射(修補細小裂縫時使用)
Crack Pattern	裂縫型式
Crack Propagation	裂縫擴散
Crack Size	裂縫尺寸
Crack Width	裂縫寬度
Cracking	裂縫，龜裂，裂紋
Cracking Moment	開裂彎矩

見下頁續…

英　　文	中　　文
Cracking Stress	裂縫應力(即混凝土之破裂模數)
Crane	吊車
Crash Truck	防撞卡車(橋樑檢測或施工時為保護工作人員免受交通撞擊而佈置之卡車)
Crash Wall	防撞牆(橋下如有鐵路或船隻經過，為保護墩身，在橋墩基礎之上作成牆狀，以保護橋墩，又稱 Collision Wall)
Crater Cracks	焊疤裂縫
Crazing	混凝土之表皮髮裂
Creep	潛變(針對混凝土結構而言)
Creep Coefficient	潛變係數
Creep Deflection	潛變變位
Creep Deformation	潛變變形
Creep Reinforcement	潛變鋼筋
Creep Strain	潛變應變
Crib Wall	框式擋土牆，格床式擋土牆
Critical Buckling Stress	臨界挫屈應力
Critical Load	臨界荷重
Critical Section	臨界斷面
Critical Stress	臨界應力

見下頁續…

英　　文	中　　文
Cross Beam (Cross Girder)	橫樑
Cross Bracing	橫隔支撐，交叉支撐
Cross Frames	橫向剛架(其作用與橫隔樑相同)
Cross Section	橫斷面，橫斷面圖
Crosstie	橫拉桿，橫拉筋
Crossover Bridge	跨越橋
Crown	路拱(亦稱橫向坡度)
Crusher	碎石機
Culverts	涵洞、涵管結構
Curb	緣石(亦稱Kerb)
Curing	混凝土養護，養生
Curing Chamber	養護室，養生室(與Curing Room同義)
Curing Compound	養護劑
Curvature	曲率
Curvature Friction	曲率摩擦(對預力結構設計而言)
Curved Box Girder	曲線箱型樑
Curved Bridge	曲線橋，曲橋
Curved Girder	曲線樑，曲樑
Curved Hybrid Girder	曲線混合樑
Curve I-Girder	曲線Ｉ型樑
Curved Tendon	曲線鋼腱

見下頁續…

英　　文	中　　文
Cut	挖方(工程圖上以C或CA表之)
Cutwater Device	橋墩臨水處防冰撞擊或減少冰、水流壓力之裝置
Cyclic Stress	循環應力
Cylinder	圓柱體模(混凝土試體)

MEMORANDOM

英　　文	中　　文
D-Cracking	"D"型裂縫
D-Value Method	D值法(AASHTO規範計算荷重橫向傳遞之方法)
Damage Evaluation	損壞估計，損壞評估
Damp Proofing Course	防濕層(通常塗抹於橋台或翼牆內側，以防止因濕而產生類似青苔之雜物)
Damping	阻尼
Damping Coefficient	阻尼係數
Damping Constant	阻尼常數
Data Base	資料庫
Datum Level	基準面
Datum Line	參考基線，基準線
Dead Crack	死裂縫(或靜止型裂縫，與Dormant Crack同義)
Dead Load	靜載重
Dead Load Factor	靜載重因數
Dead Load Moment	靜載重彎矩
Dead Load Point of Contraflexure	靜載重彎矩反曲點(此點之彎矩值為零)
Dead Load Stress	靜載重應力
Debris	流動性河川上之廢棄物、雜物等之通稱
Decay Coefficient	衰敗係數，惡化係數
Decision Sight Distance	應變視距

見下頁續…

英　　文	中　　文
Deck	橋面版
Deck Barge	台船
Deck Joint	橋面版縫
Deck Plate	床版，床鈑
Deck Reconstruction	橋面版重建，橋面版重鋪
Deck Replacement	橋面版改建
Deep Beam	深樑
Deep Foundation	深基礎(如樁基礎即屬之)
Deep Well	深水井
Defect	缺陷、瑕疵
Deflection	變位
Deformed Rebar (Deformed Bar, Deformed Reinforcement)	竹節鋼筋(日本工程界以異形鋼筋稱之)
Deformed Wires	竹節鋼絲，竹節鋼線
Degradation	河床沖刷
Degrading	降級，惡化
Degree of Compaction	夯實度
Degree of Consolidation	壓密度
Degree of Curve	曲度
Degree of Freedom	自由度
Degree of Saturation	飽和度
Dehydration	脱水

見下頁續…

英　　文	中　　文
Delamination	層隙(隱藏於混凝土內部之空隙)
Delamination Detection Machine	層隙偵測器
Delamination Test	層隙測試
Demolition	拆除(對現存結構物而言)
Density of Traffic	交通密度
Department of Transportation (DOT)	美國各州之運輸部，直屬於各州政府、聯邦政府之運輸部則以US DOT稱之
Deputy Chief Engineer	副總工程師，副總工程司
Design Change Order	變更設計(工程界以Change Order簡稱之)
Design Code	設計規範
Design Constant	設計常數
Design Details	設計細部，設計細節
Design Drawings	設計圖
Design Hourly Volume (DHV)	設計小時交通量
Design Ice Thickness	設計冰厚(針對寒帶地區之下部結構設計而言)
Design Level of Service	設計服務水準
Design Life	設計壽命，設計年限
Design Loads	設計荷重，設計載重

見下頁續…

英　　　文	中　　　文
Design Parameters	設計參數
Design Pile Capacity	設計樁強度
Design Plans	設計圖
Design Review Team	設計審核小組
Design Seismic Coefficient	設計地震係數
Design Specifications	設計規範
Design Speed	設計速率
Design Stage	設計階段
Design Strength	設計強度
Design Stress Range	設計應力範圍(針對疲勞設計而言)
Design Truck Volume (DTV)	設計卡車交通量(針對疲勞設計而言)
Design Vehicle	設計用車(例如HS20)
Design Wheel Load	設計輪重
Destructive Test (DT)	破壞性檢測，破壞性試驗
Detailed Design	細部設計(初步設計後之設計作業)
Detailed Design Consultants (DDC)	細部設計顧問
Deterioration	劣化(通常指混凝土與鋼構因齡期之增長而自然產生之老化現象)

見下頁續…

英　　　文	中　　　文
Detour	繞道，改道
Development Length	發展長度(指鋼筋而言)
Deviation Block	隔塊(通常用於後拉預力混凝土箱型樑內部)
Deviation Saddle	隔鞍座(對節塊式後拉預力混凝土橋樑而言)
Deviator	隔版
Dewatering System	抽水系統，排水系統
Diagonal	斜向結構構件之簡稱
Diagonal Crack	斜向裂縫，斜裂縫
Diagonal Member	斜構件，斜桿件
Diagonal Tension Failure	斜向受拉破壞
Diagonal Tensile Rebar	斜拉鋼筋
Diagonal Tensile Stress	斜拉應力
Diamond Saw	鑽石鋸(用於鋸、切與鑽掘混凝土之工具之一種)
Diaphragm	橫隔樑、隔版
Diaphragm Wall	連續壁
Differential Deflection	差異變位
Differential Settlement	差異沈陷，不均勻沈陷
Dike (Dyke)	擋水土堤
Dilatancy	砂土之鼓脹
Dimensional Analysis	因次分析

見下頁續…

英　　文	中　　文
DIN	德國工業標準(Deutsche Industrie Norm之簡稱)
Direct Design Method (DDM)	直接設計法
Direct Load	直接荷重
Direct Shear Test	直接剪力試驗
Disc Bearings	盤式支承
Discoloration	脫色
Disc Sander	砂輪機(用於清除鋼鐵表面之雜物)
Distance Post	里程標
Distortion	畸變(橫斷面變形之行為)
Distribution Load	分佈性荷重
Distribution Coefficient	傳遞係數，分佈係數
Distribution of Wheel Loads	輪重之傳遞與分佈
Distribution Plate	承壓版(用於後拉預力混凝土結構錨碇區)
Distribution Reinforcement	分佈鋼筋
Ditch	溝渠，邊溝
Dolphin	防撞群樁(位於水中之群樁，用以防止船隻撞擊橋台或橋墩)

見下頁續…

英　　文	中　　文
Dormant Crack	靜止型裂縫(即死裂縫，Dead Crack)
Double Check	複檢
Double-Deck Bridges	雙橋面版式橋樑
Double Friction	雙面摩擦
Double Reinforcement	複筋
Double Shear Capacity	雙剪強度(對螺栓而言)
Double T-Beam	雙T型樑
Double Wall Piers	雙牆式橋墩
Down Grade	下坡
Downstream	下游
Drafting Room	繪圖室，製圖室
Drainage	排水
Drainage Area	集水區(與Catchment Area同義)
Drainage Coefficient	排水係數
Drainage Ditch	排水溝
Drainage Pipe	排水管
Draped Tendon	彎折鋼腱
Draw Bridge	可移動式橋樑之通稱
Dredging Line	挖泥線，挖泥面
Drilling	鑽掘，鑽孔
Drip Notch	滴水凹槽
Driving Stress	樁擊應力(因打樁而引起)
Drop Hammer	落錘(打樁用)

見下頁續…

英　　　文	中　　　文
Dry Construction Method	乾式施工法
Dry Curing	乾養護
Dry-Pack Method	乾壘法(混凝土結構修補之方法之一，亦稱 Mortar-Fill Method)
Dual Axle Vehicle	雙車軸車輛
Duct	通管，套管(用於後拉預力混凝土橋樑結構)
Duct Support	套管支座
Duct Tape	套管膠帶
Ductile Fracture	延展性斷裂
Ductility	延展性
Ductility Limits	延展極限
Dumbbell Pier	啞鈴式橋墩
Dumped Riprap	拋石
Durability	耐久性
Dusting	粉化，粉末
Dye Penetrant	染色滲透劑
Dynamic Earth Pressure	動土壓力
Dynamic Effect	動力效應
Dynamic Ground Stability	地表動力穩定度
Dynamic Ice Forces	動冰力
Dynamic Load	動荷重
Dynamic Modulus of Elasticity	動彈性模數

見下頁續…

英　　　文	中　　　文
Dynamic Poisson's Ratio	動柏松比
Dynamic Testing	動力測試
Dynamic Water Pressure	動水壓

MEMORANDOM

英　　文	中　　文
Earth Embankment	土堤
Earth Excavation	土方開挖
Earth Fill	填土
Earth Pressure	土壓力
Earth Pressure at Rest	靜止土壓力
Earth Retaining Structure	擋土結構
Earth Work, Earthwork	土方工程
Earthquake	地震
Earthquake Response Spectrum	地震反應譜
Eccentric Load	偏心載重
Eccentricity	偏心
Economical Analysis	經濟分析
Eddy Current Testing	渦電流檢測法(非破壞檢測之一種)
Edge Distance	邊距
Effective Angle of Soil Friction	土壤摩擦有效角度(用於擋土牆設計)
Effective Area	有效面積
Effective Buckling Length	有效挫屈長度
Effective Compressive Modulus of Elastomer	彈體有效壓力模數
Effective Deck Span	有效版跨

見下頁續…

英　　　文	中　　　文
Effective Depth	有效深度
Effective Flange Width	有效翼版寬
Effective Footing Area	有效基腳面積
Effective Footing Length	有效基腳長度
Effective Length Factor	有效長度因數
Effective Modulus of Elasticity	有效彈性模數
Effective Moment of Inertia	有效慣性矩
Effective Overburden Pressure	有效覆土壓力
Effective Peak Acceleration (EPA)	有效尖峰加速度
Effective Prestressing	有效預力，有效預應力
Effective Size	有效粒徑
Effective Slab Width	有效版寬
Effective Span Length	有效跨距長
Effective Stiffness	有效勁度
Effective Stress	有效應力
Effective Stress Friction Angle	有效應力摩擦角(設計擋土牆時用之)
Effective Width	有效寬度

見下頁續…

英　　　文	中　　　文
Effective Width of Slab	有效版寬
Efflorescence	白華、析晶
Effringite	鈣釩石
Eigenvalue	特徵值
Elastic Bearing	彈性支承
Elastic Buckling	彈性挫屈
Elastic Deformation	彈性變形
Elastic Forces	彈性力
Elastic Foundation	彈性基礎
Elastic Limit	彈性極限
Elastic Load	彈性荷重
Elastic Range	彈性範圍
Elastic Section Modulus	彈性斷面模數
Elastic Seimic Response Coefficient	彈性地震反應係數(與地震係數同義)
Elastic Seismic Response Spectrum	彈性地震反應譜
Elastic Shortening	彈性縮短(用於預力混凝土結構)
Elastic Support	彈性支承，彈性支撐
Elastomer	彈體
Elastomeric Bearing Pad	合成橡膠支承墊

見下頁續…

英　文	中　文
Electrode	焊條，電焊條
Electro Gas Welding (EGW)	電熱氣體焊接
Electro Osmosis	電解滲流作用(用以提升中性化混凝土之酸鹼值並鈍化鋼筋)
Electro Slag Welding (ESW)	電熱熔渣焊接
Electrolyte	電解質
Electrolytic Corrosion	電解腐蝕
Element	元素
Element Stiffness Matrix	元素勁度矩陣
Elevated Highway	高架公路
Elevation	高程(簡稱EL)，立面圖(正面圖)
Elliptic Arch	橢圓拱
Elongation	伸長，延伸
Embankment	土堤，路堤
Embankment Foundation	路堤基礎
Embedment Depth	埋入深度(例如設計鋼鈑樁時便需此值)
Emulsions	乳劑
End Bearing Piles	點承樁

見下頁續…

英　　文	中　　文
End Block	端塊(用於後拉預力樑之首尾處)
End Diaphragm	端隔樑，端隔版
End Post	端柱
End Span	邊跨(多孔式橋樑兩側與橋台相鄰之跨距)
End Treatment	端點處理
Engineering Judgement	工程判斷
Environment Impact Assessment (EIA)	環境影響評估
Environmental Impact Study (EIS)	環境衝擊研究
Environmental Protection Agency (EPA)	美國政府部門之環保單位
Epicenter	震央(震源垂直投影至地表之點)
Epoxide	環氧樹脂(歐洲之稱呼)
Epoxy	環氧樹脂(亦稱Epoxy Resins)
Epoxy Bonding Coat	環氧樹脂接著塗料
Epoxy Bonding Compound	環氧樹脂黏結劑
Epoxy Cement	環氧水泥
Epoxy-Coated Rebar	環氧樹脂被覆鋼筋
Epoxy Grout	環氧漿

見下頁續…

英　　文	中　　文
Epoxy Injection	環氧樹脂注射
Epoxy Resin	環氧樹脂
Equation of Motion	運動方程式
Equilibrium	平衡
Equivalent Beam on Elastic Foundation (EBEF)	等值彈性基礎樑
Equivalent Diameter	等值直徑
Equivalent Joint Load	等值節點荷重
Equivalent Moment Factor	有效彎矩因數
Equivalent Plate Thickness	等值版厚
Equivalent Rectangular Stress Block	等值矩形應力塊(簡稱 Equivalent Stress Block)
Equivalent Static Horizontal Force	等值水平靜力
Equivalent Stress Block	等值應力塊
Equivalent Tractor Mass	等值拖車質量
Equivalent Uniform Load	等值均佈荷重
Equivalent Viscous Damping Ratio	等效黏阻尼比
Erection Analysis	架設分析
Erection Drawings	架設施工圖

見下頁續…

英　　　文	中　　　文
Erection Load	架設荷重
Erosion	侵蝕(通常指水對河川、坡地之侵蝕)
Error of Closure	閉合差
Euler Buckling Stress	尤拉挫屈應力
Excavation	開挖，挖方
Exclusive Highway	專用公路
Exit Ramp	出口匝道
Expansion Coefficient	膨脹係數
Expansion Bearing	可伸縮性支承
Expansion Joint	伸縫
Expansive Soil	膨脹性土壤
Expert System	專家系統
Expert Consulting System	專家諮詢系統
Exploration	鑽探
Exposed Reinforcement	鋼筋外露
Expressed Highway	快速公路
Express Way	快速公路
Exterior Beam	外樑
Exterior Form	外模
Exterior Girder	外樑
External Posttensioning	外置後拉預力，外部後拉預力
External Prestressing	外置預力，外部預力

見下頁續…

英　　文	中　　文
External Diaphragm	外橫隔樑，外隔版(用於兩相鄰箱型樑之間)
External Duct	外部套管，外套管
External Tendon	外置鋼腱，外部鋼腱，外鋼腱
Extrados	拱外弧面(與Back同義)
Exudation	滲膠(混凝土缺陷之一種)
Eyebar	眼桿

英　　文	中　　文
Factor of Safety (F.S.)	安全係數(亦可稱Safety Factor)
Factory	工廠
Failure Mechanism	破壞機構
Failure Surface	破壞面
Falling Weight Deflectometer (FWD)	落體測位儀
Falsework	鷹架，臨時模架
Fascia	橋樑結構之最外緣線
Fascia Girder	外樑(橋樑上部結構最外邊之大樑)
Fast-Setting	快凝，速凝
Fatigue Analysis	疲勞分析
Fatigue Category	疲勞分類(例如AASHTO規範將鋼材之疲勞分為A、B、C、D、E、E'、F與F'等類別)
Fatigue Check	疲勞檢核(針對疲勞設計而言)
Fatigue Crack	疲勞裂縫
Fatigue Crack Initiation	疲勞裂縫之初始階段
Fatigue Crack Propagation	疲勞裂縫之擴展階段
Fatigue Damage	疲勞損壞(指橋樑之損壞乃因疲勞造成)

見下頁續…

英　　文	中　　文
Fatigue Strength	疲勞強度(構件耐疲勞之能力)
Fatigue Stress Limit	疲勞應力極限
Fatigue Stress Range	疲勞應力範圍
Fatigue Truck	疲勞設計採用之卡車載重
Feasibility Study	可行性研究
Federal Emergency Management Agency (FEMA)	美國聯邦緊急事件管理署
Federal Highway Administration (FHWA)	美國聯邦公路總署，美國聯邦運輸部(US DOT)底下一個負責高速公路之專屬部門
Fender	橋墩防撞結構(如為防冰撞擊稱為Ice Guard)
Fiber Reinforced Concrete (FRC)	纖維鋼鋼筋混凝土
Fiber Reinforced Plastic (FRP)	塑鋼
Field Connection	現場施工之接頭
Field Inspection	橋樑之現場檢測
Field Splice	現場組立接合
Fill	填方(工程圖上以F或FA表之)

見下頁續…

英　　文	中　　文
Filled Joint	封閉縫(橋面版伸縮縫之一種)
Filler Plates	塡版
Fillet Weld	角焊
Filter Layer	過濾層
Final Setting	混凝土之終凝
Fine Aggregate	細骨材(例如細砂即屬之)
Fine Crack	細裂縫
Fineness Modulus (FM)	細度模數
Finger Joint	鋸齒縫，指狀縫
Finger Plate Joint	指鈑縫(又稱鋸齒鋼鈑縫)
Fire Damage	火害
Finite Difference Method	有限差分法
Finite Deformation Theory	有限變形理論
Finite Element Method	有限元素法
Fixed Arch	固定拱(與Hingeless Arch同義)
Fixed Bearing	固定式支承(即結構分析中之Hinge)
Fixed Bridge	固定式橋樑
Flame Straightening	熱燄直化，加熱矯正(鋼材直化維修法之一種)
Flange Reinforcement	翼版鋼筋

見下頁續…

英　　文	中　　文
Flanged Injection Nipple	嘴子(注射工法之灌漿注入處)
Flanges (Flange Plates)	翼版(樑之最上與最下緣部份，上緣稱為上翼版或頂版，下緣稱為下翼版或底版)
Flashlight	手電筒(橋樑檢測必備之工具之一)
Flat Duct	扁平式套管(預力混凝土套管之一種)
Flat Slab	平版
Flaws	瑕疵，缺陷
Flexibility	柔性，柔度(勁度之反義)
Flexibility Factor	柔度因數
Flexible Anchored Wall	柔性錨碇牆
Flexible Bearing	柔性支承
Flexible Sealant	柔性填縫料
Flexible Pavement	柔性路面(或稱瀝青混凝土路面)
Flexible Pier	柔性橋墩
Flexible Pile	柔性樁
Flexible Retaining Wall	柔性擋土牆
Flexural Failure	彎曲破壞
Flexural Members	受彎構件，彎曲構件
Flexural Strength	彎矩強度

見下頁續…

英　　文	中　　文
Flexure-Shear Crack	彎剪裂縫
Floating Bridge	浮橋
Floating Caisson	浮式沈箱
Floating Foundation	浮式基礎
Flood Frequency	洪峰頻率
Floor Beam	版樑(通常指與橋樑長度方向垂直之樑)
Flow Charts	流程圖
Flow Curve	流動曲線
Flow Line	流線
Flow Net	流網
Flutter	顫振(由風力引起)
Flux-Cored Arc Welding (FCAW)	包藥焊線電弧焊接
Fly Ash	飛灰
Foot Brigde	人行路橋(與Pedestrian Bridge同義)
Footing	基腳
Footing Excavation	基腳開挖
Footing Shape Factor	基腳形狀因數
Forging Steel	鍛鋼
Form Face	模板面
Form Rail	模軌
Form Traveler	模板移動架，移動式模板作業車

見下頁續…

英　　文	中　　文
Forms	模版
Foundation	基礎
Foundation Capacity	基礎承載力
Foundation Failure	基礎失敗，基礎破壞
Foundation Grillage	基礎格床
Foundation Pressure	基礎壓力
Fracture	斷裂破壞
Fracture Critical Member (FCM)	斷裂控制桿件(鋼結構構件分類之一種)
Fracture Toughness	斷裂韌度
Frame	剛架
Framing Plan	橋樑上部結構之架構平面圖
Free Body Diagram	自由體圖
Free End	自由端(可自由變位者)
Free-Standing Abutment	自立式橋台
Free Water	自由水，游離水
Freeway	高速公路
Frequency Domain	頻率域
Friction Angle	摩擦角
Friction Curvature Coefficient	摩擦曲率係數(預力結構設計時用之)
Friction Factor	摩擦因數(設計擋土牆時用之)
Friction Losses	摩擦損失(對預力結構而言)
Friction Pile	摩擦樁

見下頁續…

英　　　文	中　　　文
Friction Wobble Coefficient	摩擦搖晃係數(肇因於預力鋼腱因套管之不正確定位而產生之摩擦)
Frontage Road	側車道
Frost Action	冰凍作用(對土壤而言)
Frost Heave	凍漲
Frost Line	凍結線
Frost Zone	冰凍區
Full Height Abutment	全高式橋台
Full Scale Test	全尺度試驗，全尺寸試驗
Fully Braced Flanges	全側撐翼鈑
Functional Obsolescence	功能性退化
Fundamental Period	基本周期(橋樑耐震設計用之)
Future Load	預加荷重(目前不存在但日後可能存在之靜載重)

MEMORANDOM

英　　　文	中　　　文
Gabion	蛇籠
Galloping	馳振(用於風力分析)
Galvanic Action	電流作用
Galvanization	鍍鋅
Galvanized Steel	鍍鋅鋼
Gap	縫，縫距
Gas Metal Arc Welding (GMAW)	氣體遮護電弧焊接
Gas Porosity	氣孔(由於焊接造成)
General Accounting Office (GAO)	美國聯邦政府之主計部
General Scour	一般沖刷(或稱全面沖刷)
General Zone	廣義區，錨碇廣義區(後拉預力混凝土結構大樑端錨前方之部位)
Generalized Displacement	廣義變位
Generalized Loading	廣義荷重
Geographic Information System (GIS)	地理資訊系統(建立橋樑管理系統時，GIS具有甚為重要之功能)
Geometric Alignment	公路之幾何線形
Geometric Design	幾何設計
Geometry Control	幾何控制
Girder	大樑

見下頁續…

英　　文	中　　文
Girder Bridge (Beam Bridge)	樑式橋
Girder End	樑端
Girder Spacing	大樑間距(相鄰大樑中心線之距離)
Glass Fiber	玻璃纖維
Grade	坡度
Grade Line	公路幾何線形之縱坡線
Grade Separation	立體交叉
Grading	級配
Grading Curve	級配曲線
Grain Size	粒徑
Grain Size Analysis	粒徑分析
Gravel	礫石
Gravitational Water	重力水
Gravity Abutment	重力式橋台
Gravity Anchor	重力地錨(通常用於大型懸索橋兩端以錨碇為其主懸索)
Gravity Retaining Wall	重力式擋土牆
Grid	格樑，格子樑
Grid Analogy Method	格樑相似法(用於曲線鋼橋分析之方法之一)
Grid Deck	格床版
Grillage	格床

見下頁續…

英　　　文	中　　　文
Grillage Foundation	格床式基礎
Groove Weld	對焊
Grooving	凹槽
Gross Area	總面積
Gross Section	總斷面
Gross Vehicle Weight (GVW)	車輛總重
Gross Weight	總重(指設計時採用之車輛)
Ground Anchor	地錨
Groundline	地面線
Ground Motion Parameters	地表運動參數
Ground Surface	地表面
Ground Surface Slope	地表坡度
Ground Water Elevation	地下水位
Ground Work	地工
Groundline	地面線
Group Loading	群組載重，群組荷重
Group Pile Loading	群樁荷重
Grout Opening	灌漿口(用於後拉預力混凝土結構，與Vent同義)
Grout Tube	灌漿管(亦稱為Grout Pipe)
Grouting Port	灌漿口

見下頁續…

英　　　文	中　　　文
Guide Specifications	指引規範
Gusset Plate	繫鈑
Gutter	側溝(排水用)
Gutter Line	排水線(橋樑鋪面與護欄接觸之位置)

英　　文	中　　文
H-Beam	熱軋之H型鋼，通常作為打樁用
H-Steel Piles	H型鋼樁
Hairline Crack	毛細裂縫(與Crazing同義)
Hairpin	夾片筋(用於後拉預力混凝土結構錨碇區之小型螺旋鋼筋
Hammerhead Pier	槌頭式橋墩
Handrail	扶手欄杆
Hand Welding	手焊(與Manual Welding同義)
Hanger	吊桿，吊具
Hanger Bars	吊筋
Hardware	硬體
Hat Head Rivet	平頭式鉚釘
Haunch	托肩，漸深鈑
Headloss	水頭損失
Headwater	水頭(水流上游之來源)
Heat Treatment	熱處理
Heave Rate	土壤膨脹率
Heaving	土壤之隆起現象
Heavy Truck	大貨車
Heel	踵版(擋土牆基礎之後端)

見下頁續…

英　　文	中　　文
Helical Reinforcement	螺栓鋼筋
Hidden Cap Beam	隱藏式帽樑
High Carbon Steel	高碳鋼
High Button Rivet	凸頭式鉚釘
High Damping Rubber Bearing	高阻尼橡膠支承墊
High Level Pavement	高級路面
High Performance Concrete (HPC)	高性能混凝土
High-Pressure Clean-Air Machine	高壓空氣機(消除污物灰塵用之)
High-Strength Bolt	高強度螺栓(簡稱HS Bolt，通常指A325或更強者)
High-Strength Concrete (HSC)	高強度混凝土
High-Strength Fasteners	高強度加勁材
High-Strength Low-Alloy Steel	高強度低合金鋼
High-Strength Wires	高強度鋼線
High Tensile Bolt	高拉力螺栓
High Tensile Steel (High Tensile Strength Steel)	高拉力鋼
High Water Level (HWL)	高水位(亦有人以H.W.簡稱之)

見下頁續…

英　　文	中　　文
Highway Aesthetics	公路美學
Highway Bridges	公路橋樑
Highway Bridge Replacement and Rehabilitation Program (HBRRP)	美國FHWA之公路橋樑改建與修復計畫
Highway Capacity	公路容量
Highway Capacity Manual (HCM)	美國TRB於1994年(第三版)出版之公路容量手冊，第二版為1985年版
Highway Density	公路密度
Hinge	鉸接，鉸
Hingeless Arch	固定拱(與Fixed Arch同義)
Hollow Slab Bridge	中空版橋
Hold-Down Device	抗拉拔設施(針對上舉力而言)
Hollow Cylinder Piles	中空圓柱樁
Hollow Pier	中空墩
Hollow Pile	中空樁
Honeycomb, Honeycombing	蜂窩(RC結構物缺陷之一種)
Horizontal Alignment	公路之水平線形

見下頁續…

英　　文	中　　文
Horizontal Bearing Capacity	水平承載力
Horizontal Cracks	水平裂縫
Horizontal Curvature	水平曲率
Horizontal Curved Girder	平面曲線樑
Horizontal Offset	水平偏距
Horizontal Reinforcement Spacing	水平鋼筋間距
Horizontal Seismic Coefficient	水平地震力係數
Horizontal Shear	水平剪力
Horizontal Shear Strength	水平剪力強度
Horizontal Stiffeners	水平加勁材
Hot Cracking	熱裂縫
Hot Mechanical Straightening	熱作直化(鋼結構修理方法之一)
Hot-Poured Rubber Asphalt	熱澆橡膠瀝青(例如用於橋面版伸縮縫之處)
Hot Pouring	現場熱澆
Hot Rolling	熱軋(鋼鑄造方法之一種，與冷軋Cold Formed反義)
Hot Weather Concrete	熱天混凝土
Humidity	濕度

見下頁續…

英　　文	中　　文
Humped Bridge	背鼓式橋樑(中間比兩端高出甚多)
Hybrid Girders	混合樑(翼鈑與腹鈑由不同強度之鋼鈑組成)
Hydration	水化(水泥與水之混合作用)
Hydraulic Pile Hammer	油壓椿錘
Hydraulic Studies	水利研究
Hydrologic Analysis	水文分析
Hydrologic Engineering Center (HEC)	美國工兵團所屬之水利工程中心
Hydrostatic Pressure	靜水壓力
Hydroplanning	水翔(水積留於路面，車輛高速行駛時，其輪胎未與路面接觸之現象)
Hypocentre	震源(地震發生之源點)
Hysteresis Loop	遲滯迴圈

MEMORANDOM

英　　　文	中　　　文
I-Beam	I型樑
I-Girder	I型樑
I-Girder Bridge	I型樑橋
Ice Breaker	破冰稜
Ice Pressure	冰壓
Immediate Settlement	立即沈陷量
Impact Attenuator	碰撞緩衝設施，衝擊吸能器
Impact-Echo Method	敲擊回音法，衝擊回波法（非破壞性檢測方法之一）
Impact Effect	衝擊效應
Impact Factor	衝擊因數(加拿大Ontario之OHBOC規範以Dynamic Load Allowance，DLA稱之)
Impact Force	衝擊力
Impact Formula	衝擊公式
Impact Load	衝擊力，衝擊荷重
Importance Classification (IC)	重要性分類(橋樑耐震設計用之)
Inadequate Penetration	熔入不足(鋼結構焊接缺陷之一)
In-Depth Inspection	詳細檢測
Inclined Load	傾斜荷重

見下頁續…

英　　文	中　　文
Incomplete Fusion	熔合不良(鋼結構焊接缺陷之一)
Incremental Launching Method (ILM)	推進工法
Indirect Design Method (IDM)	間接設計法
Indirect Load	間接荷重
Inelastic Buckling	非彈性挫屈
Infinitely Long Foundation	無限長度基礎
Influence Coefficient	影響係數
Influence Depth	影響深度(通常指水對基礎之影響深度)
Influence Line	影響線
Influence Surface	影響面
Infrared Thermographic Techniques	紅外線溫度感測法(非破壞性檢測之一種)
Infrastructure	公共工程結構，公共設施
Inherent Defects	材料之固有缺陷
Initial Precompression	初始預壓
Initial Prestress	初始預力
Initial Setting	混凝土之初凝
Initial Stress	初始應力
Injection Method	注射工法，注入工法
Injection of Grout	灌漿，注漿

見下頁續…

英　　文	中　　文
Injector (Injection Gun)	注射器(混凝土維修時壓力灌漿或注射工法所用之注射器)
Inlet	入水口
Inorganic Zinc Shop Primer	無機鋅粉底漆
Inside Diameter	內徑
Inside Girder	內樑(與Interior Girder同義)
Inside Exterior Girder	內側外樑
Inspection	檢測，檢查
Inspection Access	檢測出入口
Inspection Catwalk	檢測步道
Inspection Form	檢測報表
Inspection Frequency	檢測頻率，檢查頻率
Inspection Ladder	檢測扶梯
Inspection Manual	檢測手冊，檢查手冊
Inspection Personnel	檢測人員
Inspector	檢測員
Instantaneous Strain	瞬時應變，即時應變
Integral Abutment	整體式橋台(橋台與上部結構整體構築者)
Integral Deck	整體式橋面版
Integrated Wearing Surface	整合式磨耗層

見下頁續…

英　　文	中　　文
Interaction	互制
Interaction Pattern	互制型態
Interchange	交流道
Interlayer Water	間層水
Interior Bearing	內支承(對多跨距橋樑而言，亦可以Interior Support稱之)
Interior Form	內模
Interior Girder	內樑(與Inside Girder同義)
Intermediate Diaphragm	中隔樑，橫隔樑(位於支承中間之跨距處)
Intermittent Weld	斷續焊接
Internal Diaphragm	內橫隔樑(位於箱型樑內部)
Internal Duct	內部套管，內套管
Internal Friction Angle	內摩擦角(土壤特性之一)
Internal Redundancy	內在贅餘度
Internal Stability	內在穩定度，內部穩定度
Internal Tendon	內部鋼腱，內鋼腱
International Association for Bridge and Structural Engineering (IABSE)	國際橋樑與結構工程協會

見下頁續…

英文	中文
International Bridge Conference (IBC)	國際橋樑研討會(每年六月於美國賓州匹茲堡舉行)
International Standard Organization (ISO)	國際標準組織
Interstate Highway	美國公路系統中之州際公路(相當於我國之國道高速公路)
Interstate Loading	軍用載重(與Military Loading同義)
Intrados	拱内弧面(與Soffit同義)
Inventory Rating	存檔評定(橋樑評定之一種)
Iron Bridge	鐵橋
Isolation Bearing	隔震用支承(例如LRB)
Isolation Joint	隔縫
Isolator	隔震器
Isotropic	等向性
Iteration Method	疊代法

英　　文	中　　文
Jack	千斤頂，起重機
Jacket	樁套(保護樁之外套)
Japanese Industrial Standards (JIS)	日本工業標準
Japan Road Association (JRA)	日本道路協會
Japanese Architectural Standard Specifications (JASS)	日本建築標準規範
Japan Society of Civil Engineering (JSCE)	日本土木工程學會
Jaw Crusher	齒鈑碎石機
Jig	機台
Job Site	工地
Joint	節點，縫
Joint Development	聯合開發
Joint Filler	填充料，填縫料
Joint Grouting	接縫灌漿
Joint Inspection	會檢(共同檢測並討論之)
Joint Sealant	填縫料
Joint Venture	共同承攬
Jointed Concrete Pavement	接縫式混凝土鋪面
Jointless Deck	無縫版(通常指橋面版上無伸縮縫)
Joist	小柵樑
Junction Box	排水用之匯流井

英　　　文	中　　　文
Lacing	綫材(通常是指連繫支撐構件，Bracing，之小型構件)
Lagging	曲面模(可以製造結構曲面之模板)
Lamellar Tear	層撕裂(兩層鋼材間之薄層撕裂缺陷)
Laminar Cracking	層裂(混凝土裂縫之一種)
Laminated Pad	積層墊
Laminated-Rubber Bearing	積層橡膠支承(簡稱LRB)
Lamination	夾層(鋼材之固有缺陷之一)
Landmark	地標(大型或造型優美之橋樑常是某地之地標)
Land Transportation	陸路運輸
Lane Loading	車道載重
Lane Width	車道寬度
Lap Joint	搭接頭
Large Deformation Effect	大變形效應
Large Diameter Pile	大口徑樁
Lateral Bracing	側向支撐
Lateran Clearance	側向淨距
Lateral Design Force Coefficient	橫向設計力係數

見下頁續…

英　　文	中　　文
Lateral Distributed Loads	橫向傳遞荷重
Lateral Earth Pressure	側向土壓力
Lateral Flexibility	側向柔度
Lateral Loads	側向力，側向荷重
Lateral Motion	側向運動
Lateral Reinforcement	側向鋼筋
Lateral Stability	側向穩定度
Lateral Torsional Buckling	側向扭曲挫屈
Lateral Water Pressure	側向水壓力
Latex Emulsion	橡膠乳液
Latex Modified Concrete (LMC)	乳膠改質混凝土
Launching Bearing	推進支承(推進工法採用之臨時支承)
Launching Gantry	推進機架
Launching Girder	推進樑
Launching Nose	推進鼻樑(用於推進工法，工程界以Nose簡稱之)
Layered Soil	層積土壤
Layout	放樣
Leaching	滲漏(混凝土結構缺陷之一)

見下頁續…

英　　文	中　　文
Lead Core	隔震用支承之鉛心
Lead-Extrusion Damper	鉛外擠阻尼器
Lead Rubber Bearing (LRB)	鉛心橡膠支承(用於橋樑隔減震)
Lean Concrete	貧混凝土
Leg	支腳
Legal Vehicles	法定車輛(例如設計時採用之HS20)
Length of Superelevation Runoff	超高漸變長度
Less Weldable Steel	劣焊接鋼
Level	水平，水平參考線
Level Bolt	定平用螺栓
Level of Service (LOS)	服務水準
Level Plate	水平調整鈑(通常用於鋼支承)
Lift Bridge	垂直昇降橋(可動式橋樑之一種)
Lifting Frame	吊架
Lifting Gantry	起重機架
Lifting Holes	吊孔(對預鑄節塊而言，此孔需預留)
Lifting Loop	吊環

見下頁續…

英　　文	中　　文
Lifting Reinforcement	吊筋
Lighting	照明
Lightweight Concrete	輕質混凝土
Limit Load	極限載重
Limit State	極限狀態
Line Bearing	線支承
Line Surcharge Load	線狀加載荷重
Linear Elastic Behavior	線彈性行為
Linear Friction Factor	線摩擦因數
Liquid Limit	液性限度
Liquid Penetration Testing	液滲檢測法(用於鋼橋之非破壞性檢測方法之一)
Liquification Failure	土壤之液化破壞
Live Crack	活性裂縫(與死裂縫反義，與Active Crack同義)
Live Load	活載重
Live Load Deflection	活載重變位(此為橋樑設計非常重要之檢核數據之一)
Live Load Distribution Constant	活載重傳遞常數
Live Load Factor	活載重因數
Load (Loading)	載重，荷重，荷載
Load-Balancing Method	載重平衡法(分析預力結構之方法之一)

見下頁續…

英　　文	中　　文
Load Capacity Rating	荷載能力評定
Load Carrying Capacity	荷載能力
Load Case	載重情況
Load Coefficient	載重係數，荷重係數
Load Combination Coefficient	載重組合係數
Load Cycles	載重循環，荷重循環(用於疲勞分析)
Load Equivalent Factors (LEF)	載重當量因數
Load Factor	載重因數，荷重因數
Load Factor Design (LFD)	載重因數設計(強度設計法，對橋樑設計而言)
Load Factor Rating (LFR)	載重因數評定
Load Inclination Factor	載重傾斜因數
Load Path Redundancy	荷載路徑贅餘度
Load Rating	載重評定
Load and Resistance Factor Design (LRFD)	載重與阻力因數設計(建築物與橋樑設計方法之一)
Load Test	載重試驗
Loaded Area	荷載面積
Loaded Length	荷載長度
Loading Reduction	載重折減
Loading Capacity	荷載容量，荷載強度

見下頁續…

英　　文	中　　文
Loading Combination	載重組合
Local Buckling	局部挫屈
Local Coordinate	局部座標
Local Deviations	區域偏差，局部偏差
Local Scour	局部沖刷
Local Scour Depth	局部沖刷深度
Local Transverse Moment	局部橫向彎矩
Location Plan	位置圖
Longitudinal Analysis	縱向分析
Longitudinal Bending	縱向彎矩
Longitudinal Bracing	縱向支撐
Longitudinal Continuity Posttensioning	縱向連續後拉預力
Longitudinal Crack	縱向裂縫
Longitudinal Earthquake Motion	縱向地震
Longitudinal Force	縱向力(即煞車力)
Longitudinal Grade	縱向坡度(針對道路而言，簡稱縱坡，Grade)
Longitudinal Joint	縱向縫
Longitudinal Posttensioning	縱向後拉預力
Longitudinal Reinforcement	縱向鋼筋
Longitudinal Stiffener	縱向加勁版，縱向加勁材

見下頁續…

英　　文	中　　文
Longitudinal Stiffener Coefficient	縱向加勁材係數
Longitudinal Torsion	縱向扭曲
Longitudinal Vertical Shear	縱向垂直剪力
Long Line Casting	長線澆注(見Long Line Method)
Long Line Method	長線法(預鑄式橋樑之製作方法之一，另一種為短線法，Short Line Method)
Long-Range Planning	長程規劃
Long Span Bridge	長跨距橋樑，大跨度橋樑
Long Term Deflection	長期變位
Long-Term Modulus of Elasticity	長期彈性模數
Long-Term Plan	長期規劃
Loss of Prestress	預力損失
Loose Sand	鬆砂
Low-Carbon Steel	低碳鋼
Low-Heat Cement	低熱性水泥
Low Relaxation Steel	低鬆弛度鋼材
Low-Slump Dense Concrete (LSDC)	低坍度密實混凝土
Low Temperature Zones	低溫區

見下頁續…

英　　文	中　　文
Low Water Level (LWL)	低水位(亦有人以L.W.簡稱之)
Lower Bound	下限
Lower Chord	下弦桿

英　　文	中　　文
Macadam	碎石
Macrostress	巨觀應力
Magnetic Particle Testing	磁粒檢測法(金屬非破壞性檢測之一種)
Magnification Coefficient	放大係數
Magnification Factor	放大因數
Magnification Ratio	放大比
Magnitude	地震之規模(簡稱M)
Main	管線
Main Beam	主樑
Main Girder	主樑，大樑
Main Load	主荷重，主要載重
Main Reinforcement	主鋼筋
Main Span	主跨
Main Suspension Cables	主懸索(懸索橋之主要受拉構件)
Maintenance	養護、保養
Maintenance & Protection of Traffic (MPT)	交通維持
Maintenance Work	養護工程
Mambrane Curing	覆膜養護
Manhole	人孔(箱型橋樑為了施工，檢測方便而預留之孔洞，以便人員可進出)

見下頁續…

英 文	中 文
Manning Formula	曼寧公式
Manual	手冊
Manual Welding	手焊(與Hand Welding同義)
Map Cracking (Pattern Cracking)	網狀裂縫
Masonry	圬石
Masonry Bridge	圬石橋(與Stone Bridge同義)
Masonry Plate	支承本身與混凝土墊間之鋼鈑
Mass Concrete	巨積混凝土
Mass Concrete Cracking	巨積混凝土裂縫
Mass Diagram	土積圖
Mat Foundation	筏式基礎
Mathematical Model	數學模式
Maturity Method	成熟度檢測法(非破壞檢測之一種)
Maximum Design Load	最大設計載重(AASHTO之LFD橋樑設計方法中必須考量之一種載重情況)
Maximum Ground Acceleration	最大地表加速度
Maximum Possible Scour Depth	最大可能沖刷深度

見下頁續…

英　　　文	中　　　文
Maximum Resisting Force	最大阻力，最大抗力
Mean Annual Relative Humidity	年平均相對濕度
Mean Shrinkage Strain	平均乾縮應變
Meandering River	曲折河川
Mean Water Level (MWL)	平均水位
Mechanical Connection	機械式接頭(例如兩條鋼筋之間以續接器相接)
Mechanically Stabilized Earth (MSE)	機械式穩定土壤
Median Strip	中央分隔帶(簡稱Median)
Member	構件，桿件
Member Stiffness Matrix	構件勁度矩陣
Membrane Curing	薄膜養生
Metal Corrosion	金屬腐蝕
Metal Duct	金屬套管
Microstress	微觀應力
Middle Coat Painting	中塗塗裝
Midspan	橋樑跨距之中點
Mild Steel	軟鋼
Military Loading	軍用載重(美國之工程界亦稱為Interstate Loading)
Minimum Reinforcement	最小鋼筋量

見下頁續…

英　　文	中　　文
Mixed Torsion	混合扭曲(即純扭曲與翹曲扭曲之和)
Mobile Scaffold	平台型檢測車
Modal Test	模型試驗
Mode Shape	模態
Modification Factor	修正因數
Modified Bearing Capacity Factor	修正承載力因數
Modified Foundation Seismic Force	修正基礎地震力
Modulus of Deformation	變形模數
Modulus of Elasticity	彈性模數，楊氏模數
Modulus of Rigidity	剛度模數
Modulus of Rupture	破裂模數
Modulus Ratio	模數比
Moisture Content	含水量(亦稱Water Content)
Moisture Control	濕度控制
Moment	彎矩(與Bending同義)
Moment Capacity	抗彎強度
Moment Connection	抗彎接頭
Moment Diagram	彎矩圖
Moment Distribution	彎矩分配
Moment Envelope	彎矩包絡線
Moment of Inertia	慣性矩

見下頁續…

英　　文	中　　文
Moment Redistribution	彎矩重新分配
Monitoring	監控
Mortar	水泥砂漿
Movable Bearing	可動式支承
Movable Bridge	可動式橋樑
Moving Load	移動式荷重(移動中之活載重)
Mudball	泥球(混凝土表面缺陷之一種)
Multi-Cell Box Girder	多室箱型樑
Multi-Cell Box Girder	多室箱型樑
Multi-Centered Arch	多心拱
Multi-Column Bent	多柱式橋墩
Multi-Deck Bridge	多層式橋樑
Multidimensional Posttensioning	多向性後拉預力

MEMORANDOM

英　　　文	中　　　文
Narrow Bridge	狹橋
National Association of Corrosion Engineers (NACE)	國際防蝕工程師協會
National Bridge Inventory (NBI)	全國橋樑檔案
National Bridge Inspection (NBI)	美國之全國橋樑檢測
National Bridge Inspection Standards (NBIS)	美國之全國橋樑檢測標準
National Bureau of Standards (NBS)	美國國家標準局(位於美國馬里蘭州)
National Cooperative Highway Research Program (NCHRP)	美國公路合作研究組織群
National Environmental Policy Act (NEPA)	國家環境政策法案(美國於1969年提出)
National Highway Institute (NHI)	美國國家高速公路學會
National Institute for Certification in Engineering Technologies (NICET)	美國工程技術檢定學會

見下頁續…

英　　文	中　　文
National Institute of Standards and Testing (NIST)	美國國家標準與試驗學會（創於 1980年）
National Science Foundation (NSF)	美國之國家科學基金會
Natural Frequency	自然頻率
Natural Period	自然周期
Natural Rubber Bearing	天然橡膠支承
Natural Vibration	自然振動
Nature Coordinate	自然座標
Necking	頸縮(構件伸長而斷面減小之現象)
Negative Bending	負彎矩
Negative Moment	負彎矩
Negative Moment Region (NMR)	負彎矩區(亦可稱 Negative Moement Range)
Negative Moment Reinforcement	負彎矩鋼筋
Negative Skin Friction	負表皮摩擦
Neoprene	合成橡膠
Net Area	淨面積(用於受拉構件設計)
Neutral Axis (N.A.)	中立軸(對彎矩分析而言)

見下頁續…

英文	中文
Neutral Surface	中立面(中立軸沿構件長度形成之面)
New Jersey Concrete Barrier	紐澤西混凝土護欄(簡稱Jersey Barrier)
Nodal Force	節點力
Nominal Dimension	標定尺寸
Nominal Diameter of Bars	鋼筋標定直徑
Nominal Shear Strength	標定剪力強度
Nominal Strength	名義強度，標定強度(理論上之強度)
Nominal Torsion Resistance	標定抗扭力
Nondestructive Examination (或 Evaluation)	非破壞性檢測(簡稱NDE)
Nondestructive Inspection	非破壞性檢測(簡稱NDI)
Nondestructive Testing (NDT)	非破壞性檢測
Nongravity Cantilevered Walls	非重力式懸臂牆
Nonlinear Analysis	非線性分析
Nonlinear Finite Elements	非線性有限元素

見下頁續…

英　　文	中　　文
Non-Prestressed Reinforcement	非預力鋼筋
Non-Shrink Grout	非伸縮性砂漿
Nonstructural Crack	非結構裂縫
Nonstructural Members	非結構功能之構件(例如排水管、照明設備等等)
Nonsymmetrical Bending	非對稱彎矩
Nonuniform Torsion	不均勻扭曲(與Warping Torsion同義)
Normal Crown (NC)	正常路拱
Normal Stress	正向應力(與橫斷面垂直之應力)
Normally Consolidated Clay	正常壓密黏土
North Arrow	指北記號
Nose, Nose Beam	鼻樑(用於推進工法，亦可稱為Launching Nose)
Notch	缺口，凹口
Notch Effect	凹口效應(指斷面突變而產生之應力集中現象)
Nozzle Sand Blasting	噴砂(清除鋼鐵表面異物、雜質之方法之一)
Nut	螺栓帽

英　　文	中　　文
Observation Well	水位觀測井
Offset	偏距
Offshore Structure	濱海(或近海)結構物
One-Way Slab	單向版
Ontario Highway Bridge Design Code (OHBDC)	加拿大之橋樑設計規範(其最新版為1992年之第3版)
Open Abutment	開放式橋台
Open Caisson	開口沈箱
Open Joint	開口縫(位於橋面版上)
Open Section	開口式斷面，開放式斷面
Open Steel Grid	開口式鋼格床
Operating Rating	運作評定(橋樑評定方式之一)
Operating Speed	可運行速率
Optical Crack Gauge	光學裂縫量測計
Organic Corrosion Inhibiting Admixture (OCIA)	有機腐蝕抑制劑
Organic Soil	有機土壤
Organization for Economic Cooperation and Development (OECD)	經濟合作與開發組織(其總部設於歐洲之巴黎)
Original Bed	原始河床

見下頁續…

英　　　文	中　　　文
Orthogonal Seismic Forces	正交異向地震力
Orthotropic Steel Plate	正交異向性鋼鈑
Out-of-Plate Bending	面外彎矩
Outlet	出水口
Outlet Pipe	放流管
Outside Diameter	外徑
Outside Exterior Girder	外側外樑
Overall Priority Index	總體優選指標
Overall Stability	總體穩定度
Overburden	覆土壓力
Overconsolidated Clay	過壓密黏土
Overhang	懸伸，懸臂
Overlay	覆蓋層，亦可作為橋面版磨耗層之簡稱
Overloading (Overload)	超載
Overpass	跨越
Overpass Bridge	跨越橋
Overreinforced Concrete	超筋混凝土
Oversize Vehicle	超尺寸車輛
Overstressing	應力超過
Overturning Forces	傾覆力
Overturning Moment	傾覆彎矩

見下頁續…

英　　文	中　　文
Overturning Stability	傾覆穩定性(用於擋土牆與橋台之設計)
Overweight Vehicle	超重車輛
Oxidation	氧化

英　　　文	中　　　文
Pachometer (Cover Meter)	混凝土保護層量測計
Paint Film Gauge	漆厚量測計
Paint Wrinkling	油漆表面之皺紋
Painting	油漆，塗裝
Parabolic Arch	拋物線拱
Parallel Structure	並連式結構
Parapet (Wall)	胸牆，女兒牆
Parking Sight Distance	停車視距
Partial Prestressing	部份預力
Partial Reconstruction	整建，部份重建
Partially Bonded	部份黏結式鋼腱
Partially Saturated Soil	部份飽和土
Particular Load	特殊荷重
Passive Earth Pressure	被動土壓力
Passive Film	鈍態膜(鋼筋與混凝土結合之後在鋼筋表面形成之保護膜)
Passive Resistance Factor	被動抗力因數(設計擋土牆時用之)
Passive Oxide Layer	鈍化氧化膜或簡稱鈍化膜(與上述之Passive Film同義)
Patch	填補(修補混凝土)
Pavement	舖面，路面
Pavement Crown	路拱

見下頁續…

英　　　文	中　　　文
Pavement Drainage	路面排水
Pavement Edge Line	路面邊線
Pavement Management System (PMS)	鋪面管理系統
Pay Limit	付款極限(通常必須標示於橋樑設計圖中，使包商於估價與施工時有參考之依據)
Peak Acceleration	尖峰加速度(用於橋樑耐震設計)
Peak Ground Acceleration	尖峰地表加速度
Pedestal	支承座墊(通常為混凝土製，其上為支承)
Pedestal Footing	單腳式基腳
Pedestrian Bridges	人行橋(Foot Bridge與同義)
Pedestrian Crossing Bridge	人行跨越橋
Pedcstrian Railing	行人手扶欄杆
Peeling	脫皮(指混凝土表面薄片狀之脫落現象)
Penetration Resistance Method (PRM)	貫入阻力法(非破壞檢測之一種，通常用於測試混凝土之表面強度)

見下頁續…

英　　文	中　　文
Perforated Pipe	洞管(管之表面具有孔洞以利排水)
Performance	績效
Permanent Bearing	永久性支承
Permanent Bridge	永久性橋樑
Permanent Deformation	永久變形
Permanent Loads	永久性荷重
Permeability	滲透性
Permeability Coefficient	透水係數
Permit Vehicles	允許性車輛
Pervious Backfill	透水性背填土
Phenolphthalein	酚太(可作檢測混凝土碳酸化之指示劑)
Phthalic Resin Coating	酸樹脂塗料
Phthalic Resin Painting	酸樹脂塗料
Pier	橋墩
Pier Cap	墩帽(可簡稱Cap)
Pier Column	墩柱
Pier Design Forces	橋墩設計力
Pier Diaphragm	墩頂橫隔樑，墩頂隔版
Pier Monitoring	橋墩監測
Pier Nose	墩鼻

見下頁續…

英　　　文	中　　　文
Pier Segment	墩頂節塊(節塊式橋樑在橋墩正上方與上部結構結合之節塊)
Pier Shaft	橋墩主壁體
Pier Stem	墩牆
Pier Wall	墩牆
Pile	樁，基樁
Pile Bent	樁排(橋墩型式之一)
Pile Bent Abutment	樁排式橋台
Pile Cap	樁帽(位於樁基礎之最上部份)
Pile Cut-off	樁切點
Pile Diameter	樁徑
Pile Driver	打樁機
Pile Footing	樁基腳
Pile Foundation	樁基礎
Pile Group	群樁
Pile Lugs	樁柄
Pile Pier	樁排式橋墩(與Pile Bent同義)
Pile-Supported Abutment	樁基橋台
Piling	樁，群樁
Pin	樞接，鎖針
Pinholes	針孔(油漆塗裝缺陷之一)
Pipe	管

見下頁續…

英　　文	中　　文
Pipe Arch	管拱
Pipe Culvert	管涵
Pitted Surface	表面凹洞
Pitting	窪坑(混凝土缺陷之一種)
Placing	混凝土之澆注、澆置
Plain Concrete	素混凝土(即無筋混凝土)
Plain Reinforcement	光滑鋼筋，滑面鋼筋
Plan	平面圖
Plane Strain	平面應變
Plane Stress	平面應力
Planimeter	求積儀
Planning Stage	規劃階段(亦稱Conceptual Statge)
Plastic Buckling	塑性挫屈
Plastic Clay	塑性黏土
Plastic Cracking	塑性裂縫(混凝土結構物之早期裂縫之一)
Plastic Deformation	塑性變形
Plastic Design	塑性設計
Plastic Hinge	塑性鉸
Plastic Moment	塑性彎矩
Plastic Pipe	塑膠管
Plastic Section Modulus	塑性斷面模數

見下頁續…

英　　　文	中　　　文
Plastic Settlement Crack	塑性沈陷裂縫
Plastic Shrinkage Cracks	塑性乾縮裂縫
Plastic Stress Distribution	塑性應力分佈
Plasticity Index (PI)	塑性指數
Plate Bearing	鈑式支承
Plate Bearing Test	平版載重試驗
Plate Girder	鈑樑(利用鋼鈑焊接而成)
Plumb Bob	鉛垂球
Plumb Line	鉛垂線
Pocket of Compressed Air	壓縮空氣袋(由套管兩端灌漿時易產生之)
Point Surcharge Load	點加載荷重
Polyethylene Duct	聚乙烯套管
Polymer Concrete (PC)	聚合物混凝土
Polymer Impregnated Concrete (PIC)	聚合物灌注混凝土
Polymeric Reinforcement	聚合物鋼筋
Polyurethane	聚胺樹脂
Polyvinylchloride Pipe	PVC管(排水用)

見下頁續…

英　　　文	中　　　文
Pontoon Bridge	浮橋(與Floating Bridge、Bateau Bridge同義)
Poor Ground	不良地盤，軟弱地盤
Popout	爆開(混凝土缺陷之一種)
Pore Water	孔隙水
Pore Water Pressure	孔隙水壓
Porosity	孔隙率(對土壤而言)
Portable Bridge	可攜帶式橋樑(構件可拆卸與運輸，並在他處組裝之橋樑)
Portland Cement	波特蘭水泥
Portland Cement Association (PCA)	波特蘭水泥協會
Positive Contact Pressure	正接觸壓力
Positive Moment Region (PMR)	正彎矩區(亦可稱Positive Moment Range)
Positive Moment Reinforcement	正彎矩鋼筋
Post-Stressing	後拉預力
Posttensioning	預力施工之後拉法(工程界以PT簡稱之)
Posttensioning Anchorages	後拉預力錨碇
Posttensioning Bars	後拉預力鋼棒

見下頁續…

英　　　文	中　　　文
Posttensioning Duct	後拉預力套管
Posttensioning Force	後拉預力
Posted Bridge	設限之橋樑(限速或限重)
Pot Bearing	盤式支承(具有多方向旋轉之功能)
Pouring Sequence	混凝土之澆灌順序
Pot Holes	混凝土表面之大型孔洞
Precast	預鑄
Precast Concrete Girder	預鑄混凝土樑
Precast Concrete Piles	預鑄混凝土樁
Precast Girder	預鑄樑
Precast Segment	預鑄節塊(針對混凝土橋樑而言)
Precast Segment Bridges	預鑄節塊式橋樑
Precast Slab	預鑄版
Precompressed Zone	預壓區(針對預力混凝土結構而言)
Prefabricated Rebar Cage	預製鋼筋籠
Prefricated Member	預製構件
Preliminary Design	初步設計
Preloading	預壓，預載
Prepacked Concrete	預疊混凝土

見下頁續…

英　　文	中　　文
Prepack Method	預壘法(橋樑維修、修補之方法之一，亦稱Preplaced Aggregate Method)
	鋼鐵表面處理度
Preparation Grade	預設
Pre-Set	壓力灌漿
Pressure Grouting	預力鋼棒
Prestress Bar	預力鋼索
Prestress Cable	預力損失
Prestress Loss	預力混凝土
Prestressed Concrete (PC)	預力混凝土樑
Prestressed Concrete Girder	美國之預力混凝土學會
Prestressed Concrete Institute (PCI)	預力混凝土樁
Prestressed Concrete Pile	預力鋼筋混凝土
Prestressed Reinforced Concrete (PRC)	預力鋼結構
Prestressed Steel Structure	預力鋼腱
Prestressed Tendon	預力施工之先拉法

見下頁續…

英　　　文	中　　　文
Preumatic Caisson	壓氣沈箱(封口式沈箱之一種)
Priliminary Design	初步設計
Primary Bending Moment	主彎矩
Primary Consolidation	主要壓密量
Primary Flexural Reinforcement	主要受彎鋼筋
Primary Grout	主漿(地錨系統最後端，其體積較大之部份)
Primary Loads	主要荷重(與Principal Loads同義)
Primary Members	主要構件(與Principal Members同義)
Primary Moment	主要彎矩，主彎矩
Primary Settlement	主要沈陷量
Primary Stress	主應力
Primer	底漆(指油漆或表面塗裝之最內層)
Priming Coat	鋼鐵構件之第一層底漆(亦稱Shop Coat或Base Coat)
Principal Stress	主應力
Priority Index	優選指標
Priority Ranking	優選排序
Profile	道路之縱剖面

見下頁續…

英　　　文	中　　　文
Progressive Placement Construction	連續佈設施工法
Project Engineer	計畫工程師，專案工程師
Project Manager	計畫經理，專案經理
Project Management	專案管理，計畫管理
Protractor	傾斜量測計(通常用於量測支承之傾斜度)
Prying Tension	槓桿拉力(通常發生於受拉之接頭)
PT Tendon	後拉預力鋼腱(PT為Posttensioning之簡稱)
Pulley	滑車
Pullout Capacity	拉拔強度
Pullout Testing Method	拉拔試驗(非破壞性檢測之一種)
Punching Index	剪穿指數(基礎設計用之)
Punching Shear	剪穿力
Pure Torsion	純扭曲(與Saint-Venant Torsion同義)
Push-Out Construction	推進工法(與Incremental Launching Method同義)
Pylon	橋塔(斜張橋與懸索橋使用之塔式橋墩，或稱Tower-Type Pier)

英　　文	中　　文
Rader Penetrating Method	雷達波檢測法(橋梁非破壞檢測常用方法之一)
Radioactive Methods	輻射法(非破壞檢測之一種)
Radiographic Testing (RT)	射線照相法(非破壞檢測之一種)
Radius of Curvature	曲率半徑
Radius of Gyration	旋轉半徑，迴轉半徑
Rail Car	軌道運輸車(軌道乃是施工時臨時設置者)
Rail Girder	軌樑
Railing	欄杆
Railway Bridge	鐵路橋
Rainbow Arch	虹式拱
Rainfall Duration	降雨延時
Rainfall Frequency	降雨頻率
Rainfall Intensity	降雨強度
Ramp	公路系統之匝道
Random Cracks	隨機型裂縫
Random Vibration	散漫振動
Rating	評定(將現存橋樑定出等級或狀況之作法)
Rating Equation	評定方程式
Rating Factor (RF)	評定因數

見下頁續…

英　　　文	中　　　文
Rating Vehicle	評定車輛(例如AASHTO之Type 3 Unit，Type 3-S2 Unit與Type 3-3 Unit均屬之)
Reaction	反力(反作用力之簡稱)
Reaction Wall	反力牆
Ready-Mix Truck	預拌混凝土卡車
Rebar	鋼筋，鋼條
Rebar Cage	鋼筋籠
Rebound Hammer Method	反彈鎚法，衝鎚法(非破壞性檢測之一種)
Recess Pocket	端錨凹口，端錨凹槽(用於後拉預力混凝土結構)
Reconstruction	重建
Recyclability	再生性(材料重複使用之能力，例如鋼材之再生性甚佳)
Reduction Coefficient	折減係數
Reduction Factor	折減因數
Redundancy	贅餘度
Redundant Force	贅餘力
Redundant Load	贅餘載重
Redundant Structure	贅餘結構
Regular-Wingwall Abutment	普通翼型橋台

見下頁續…

英　　文	中　　文
Rehabilitation	修復(工程上簡稱Rehab.)
Rehabilitation Report	修復報告書
Reinforced Concrete (RC)	鋼筋混凝土
Reinforced Concrete Box Culverts	鋼筋混凝土箱涵
Reinforced Concrete Pavement	鋼筋混凝土鋪面
Reinforced Concrete Pipe	鋼筋混凝土管
Reinforced Earth	加勁土
Reinforced Earth Abutment	加勁土式橋台
Reinforcement	鋼筋，亦稱Steel Bar
Reinforcement Cage	鋼筋籠
Reinforcement Cover	鋼筋覆蓋層(工程界以Cover簡稱之)
Reinforcement Ratio	鋼筋比
Reinforcement Spacing	鋼筋間距
Reinforcing	補強
Reinforcing Bar	鋼筋
Reinforcing Pile	加強樁
Reinforcing Steel	鋼筋
Relative Consistency	相對稠度
Relative Density	相對密度

見下頁續…

英　　　文	中　　　文
Relative Displacement	相對變位
Relative Humidity	相對濕度
Relative Profile Gradient	超高漸變率
Relative Stiffness	相對勁度
Relaxation of Tendon Stress	鋼腱應力之鬆弛
Reliability	可靠性
Remaining Life	剩餘壽命
Remaining Section	剩餘斷面(鋼構件銹蝕後所剩之斷面)
Repair	修理，修復
Repeating Loads (Repetitive Loads)	重複性荷重
Replacement	改建
Repose Angle	安息角，靜止角(針對砂土而言)
Research Council on Structural Connections (RCSC)	美國之結構接合研究學會
Residual Deformation	殘餘變形
Residual Stress	殘餘應力，殘留應力
Resistance Factor	抗力因數(即強度折減因數，Capacity Reduction Factor)

見下頁續…

英　　　文	中　　　文
Resistivity Method	電阻檢測法(橋樑非破壞檢測方法之一)
Resonance	共振
Resonant Frequency Method	共振頻率法(非破壞性檢測方法之一)
Response Modification Factor	反應修正因數(橋樑耐震設計用之，簡稱R Factor)
Restoring Force	回復力，復原力
Restrainer Cable	束限索
Restrainer Rod	束限桿
Resultant	合力
Resurfacing	混凝土表面之重新修飾
Retaining Wall	擋土牆
Retarder	緩凝劑(用於混凝土結構)
Retrofitting	補強修復
Revealing	瀝青磨耗層之表面脫皮
Reverse Circulation Pipe	反循環樁
Ribbed Footing	肋床式基腳
Rib Shortening	肋樑縮短
Riding Quality	行車品質
Rigging	工作台(橋樑檢測常用之)
Right Bridge	直交式橋樑(與Squared Bridge同義)
Right-of-Way (ROW)	路權

見下頁續…

英　　　文	中　　　文
Right-of Way Width	路權寬度
Rigid Bearing	剛性支承
Rigid Body	剛體
Rigid Connection	剛性接頭(剛接)
Rigid Culverts	剛性管涵
Rigid Duct	剛性套管
Rigid Filler	剛性填縫料(例如修理死裂縫用之)
Rigid Frame	剛架
Rigid Frame Bridge	剛架式橋樑
Rigid Gravity Walls	剛性重力式牆
Rigid Joint	剛接點
Rigid Pavement	剛性路面(即鋼筋混凝土路面)
Rigid Retaining Wall	剛性擋土牆
Rigidity	剛度
Riprap	堆石，拋石(用於保護土堤或下部結構)
Risk Analysis	風險分析
Rivet	鉚釘
Riveted Connection	鉚接(鉚釘接合，鉚釘接頭)
Roadbed	路床，路基
Road Rater	道路評定儀
Roadway Alignment	公路線形
Roadway Crown	路冠

見下頁續…

英　　文	中　　文
Robotic Inspection Method	機器人檢測法
Rocker Bearing	搖軸支承
Rock Mass Modulus	岩石質量模數
Rock Strata	岩層
Rolled Beams	軋型樑(指熱軋鋼而言)
Rolled Steel	型鋼，熱軋型鋼
Roller Bearing	滾軸支承
Root Cracks	根部裂縫(焊接缺陷之一種)
Rotation	旋轉
Rotation Capacity Tests	旋轉能力測試
Roughness	粗糙度
Routing Analysis	路徑分析
Routing Inspection	日常檢測
Rubber Bearing	橡膠支承
Rubbing	磨光
Running Speed	行車速率
Runoff Coefficient	逕流係數
Rust Grade	鋼鐵銹蝕度
Rust-Preventing Paint	防銹漆
Rust-Resisting Paint	防銹漆
Rust Stains	銹斑
Rutting	車轍(路面缺陷型式之一種)

MEMORANDOM

英　　文	中　　文
Saddle Hinge	鞍座式支承(混凝土支承之一種)
Safe Load	安全荷重(結構物能有效承載者)
Safe Load-Carrying Capacity	安全承載能力
Safety Belt	安全帶(在某一高度以上從事橋樑施工或檢測用之)
Safety Inspection	安全性檢測，安全檢測
Sag	下垂(通常指纜索結構因自重而下垂之效應)
Sag Ratio	垂量比
Salt Attacks	鹽害
Sampler	採樣器
Sand	砂，矽砂(噴砂法清除鋼鐵表面雜物用之)
Sandblasting	噴砂
Sand Pile	砂樁
Saw Cut	鋸切
Sawed Joint	鋸縫
Scaffold, Scaffolding	鷹架，施工支撐架
Scaling	鱗片剝落(混凝土缺陷之一)
Scanning Electronic Microscope	掃描電子顯微鏡

見下頁續…

英　　文	中　　文
Schmidt Rebound Hammer	施密特反彈錘(非破壞性檢測中反彈錘試驗法常用之工具)
Scour	沖刷
Scour Protection	沖刷保護
Scour Protective System	沖刷保護系統
Screed	刮板(刮平濕軟混凝土之用)
Screeding Board	刮板
Scupper	排水管
Sealing Materials	封面材料
Sealant	填縫料
Seals	填封料
Seal Welds	封焊
Seam	接合線，縫合線
Seam Strength	接合強度(用於管涵之設計)
Seat of Approach Slab	進橋版座
Secant Modulus of Elasticity	切線彈性模數
Secondary Consolidation	次要壓密量
Secondary Loads	副荷重，次要荷重
Secondary Members	次要構件，副構件
Secondary Moment	次要彎矩，副彎矩
Secondary Stress	副應力，次應力

見下頁續…

英　　文	中　　文
Section Loss	斷面損失
Section Modulus	斷面模數
Sectional Properties	斷面性質
Seepage	滲流(通常指水滲透進入土壤內之緩慢流動現象)
Segment	節塊
Segment Casting	節塊澆鑄
Segment Erection	節塊架設
Segment Joint	節塊接縫
Segmental Box Girder	節塊式箱型樑
Segmental Bridges	節塊式橋樑
Segmental Concrete Bridge	節塊式混凝土橋樑
Seismic Coefficient	地震係數
Seismic Coefficient Method	地震係數法
Seismic Force	地震力
Seismic Isolation	隔震
Seismic Loading Intensity	地震荷重強度
Seismic Performance Category (SPC)	地震績效類別(橋樑耐震設計用之)
Seismic Zone	地震分區(震區)
Self-Anchoring Suspension Bridge	自錨式懸索橋

見下頁續…

英　　文	中　　文
Self-Excited Oscillations	自勵振動
Selfweight	自重
Semi-Automatic Welding	半自動焊接
Semi-Gravity Abutment	半重力式橋台
Semi-Gravity Retaining Wall	半重力式擋土牆
Semi-Stub Abutment	半短塊式橋台
Senior Engineer	資深工程師
Serious Structure	串連式結構
Severe Scalling	嚴重剝落(混凝土結構物缺陷之一種)
Service Life	服務年限
Service Load	服務荷重(或稱工作荷重，Working Load)
Service Load Design	服務載重設計
Service Load Stresses	服務載重應力
Serviceability	服務性
Setting Period (Setting Time)	混凝土之凝結時間
Settlement	沈陷
Settlement Detector	沈陷偵測器
Settlement Gauge	沈陷計
Settlement Point	沈陷觀測點
Settlement Ratio	沈陷比

見下頁續…

英　　文	中　　文
Settlement Slab	沈陷版
Shaft	主壁體(橋台或橋墩之主要部份)
Shallow Foundation	淺基礎
Shape Factor	形狀因數(例如設計合成橡膠支承墊便需此值)
Shape Function	形狀函數
She Bolt	陰螺
Shear	剪力
Shear Capacity	剪力強度
Shear Center	剪力中心
Shear Coefficient	剪力係數
Shear Connector	剪力連接器
Shear Envelope	剪力包絡線
Shear Failure Plane	剪力破壞面
Shear Friction	剪力摩擦，剪摩擦(鋼筋混凝土結構設計重要考量之一)
Shear Friction Reinforcement	剪力摩擦鋼筋
Shear Key	剪力榫
Shear Lag	剪力遲滯
Shear Modulus	剪力模數

見下頁續…

英　　文	中　　文
Shear Modulus of Elasticity	剪彈性模數(與Modulus of Transverse Elasticity同義)
Shear Reinforcement	剪力鋼筋
Shear Strain	剪應變
Shear Strength	剪力強度，抗剪強度
Shear Stress	剪應力
Shear Stress Coefficient	剪應力係數
Shear Studs	剪力釘
Shear-Tension Cracks	剪拉裂縫
Shear-Tension Failure	剪拉破壞
Shear Wall	剪力牆
Shearing Test	剪斷試驗
Sheath, Sheathing	套管(預力結構或地錨系統用之)
Sheath Duct	套管(用於預力混凝土結構)
Sheet Pile	鈑樁(Steel Sheet Pile，鋼鈑樁之簡稱)
Sheet Pile Cofferdam	鋼鈑樁圍堰
Shielded Metal Arc Welding (SMAW)	被覆電弧焊接
Shim Plate	加撐版
Shop Drawing	施工詳圖
Shop Fabricated	工廠預製

見下頁續…

英　　　文	中　　　文
Shop Primer	防銹底漆(指鋼材防銹處理之第一層防銹底漆)
Short Line Casting	短線澆注(見Short Line Method)
Short Line Method	短線法(預鑄式橋樑之製作方式之一，另一種為長線法，Long Line Method)
Shot	鋼球(利用噴射原理對鋼材除銹使用之，另兩種為利用砂，Sand，與粗粒子，Grit)
Shotcrete	噴凝土
Shoulder	路肩
Shrinkage	乾縮(由於混凝土硬化乾燥所致)
Shrinkage Crack	乾縮裂縫
Shrinkage Effect	乾縮效應
Shrinkage Limit	乾縮限度
Shrinkage Reinforcement	乾縮鋼筋
Shrinkage Strain	乾縮應變
Shrinkage Stresses	乾縮應力
Side Form	外模，邊模
Side Resistance	側向阻力
Side View	側視圖

見下頁續…

英　　文	中　　文
Sidewalk	人行步道
Sidewalk Floor	人行道版
Sieve Analysis	篩分析
Sight Distance	視距(公路幾何設計必須考量之因素之一)
Silty Soil	沈泥土壤
Simple Beam	簡支樑
Simple Shear Test	單剪試驗
Simple Span	簡支跨距，單跨，單孔
Simple Support	簡支承，簡支撐
Simply Supported Bridge	簡支橋樑
Single Friction	單面摩擦
Single Shear Capacity	單剪強度(對螺栓而言)
Site Coefficient	橋址係數(橋樑耐震設計用之)
Site Effects	橋址效應(橋址地質對橋樑耐震之影響)
Size Factor	尺寸因數
Skewed Angle	斜角，斜交角
Skewed Bridge	斜式橋(交角非90°者)
Skewed Crossing	斜式交叉
Skid Resistance	抗滑性(針對橋面版或路面而言)
Slab Bridge	版橋
Slag Entrapment	夾渣(鋼材焊接缺陷之一種)

見下頁續…

英　　文	中　　文
Slenderness Effect	細長效應(受壓構件必須考量之因素之一)
Slenderness Ratio	細長比
Sliding	滑移，滑動
Sliding Failure	擋土結構或邊坡之滑動破壞
Sliding Plate Bearing	滑鈑支承
Sliding Plate Joint	滑鈑縫
Sling	吊索
Slip Coefficient	滑動係數(Slip-Critical Joint之設計用之)
Slip Critical Connection	抗滑式接頭
Slipperiness	滑動性(橋面版設計時必須考量其表面具足夠之抗滑性)
Slope	邊坡，斜率
Slope Failure	邊坡破壞，邊坡失敗
Slope Paving	邊坡舖面
Slope Protection	邊坡保護
Slope Stability	邊坡穩定
Slump	混凝土之坍度
Slurry Wall	連續壁
Small Spall	小型剝離
Smooth Bar	光面鋼棒，光面鋼筋
Snopper	多臂桿型橋樑檢測車

見下頁續…

英　　文	中　　文
Soil Erosion	土壤侵蝕
Soil Friction Angle	土壤摩擦角
Soil Modulus	土壤模數
Soil Parameters	土壤參數
Soil-Pile-Foundation Interaction	土壤－樁－基礎互制
Soil Profile	土壤剖面
Soil Strata	土層
Soil-Structure Interaction	土壤結構互制
Soil-Structure Interaction Factor	土壤結構互制因數
Sole Plate	支承水平調整鈑
Solid Piles	實心樁
Solid Piers	實心墩
Solid-Wall Pier	實心牆式橋墩
Solvent Cleaning	溶劑洗淨法(鋼鐵表面銹蝕清除方法之一種)
Sound Detection Method	音波檢測法(非破壞檢測之一種)
Sounding	聲測(用於檢測混凝土是否密實)
Soundness	健全度
Spacer	間隔物

見下頁續…

英　　文	中　　文
Spalling	剝離(指混凝土成片塊地脫落)
Span	跨距，跨徑，跨度
Span-by-Span Construction	逐跨施工法
Span-by-Span Erection	逐跨架設
Span Length	跨距長度，跨徑長度
Spandrel	拱腔
Special Loading Conditions	特殊荷重狀況
Special Project	專案計畫
Special Segment	特殊節塊
Specifications	規範
Spherical Bearing	球面支承
Spider	橋樑檢測時之高空吊纜
Spill-Through Abutment	直穿式橋台
Spiral	螺旋曲線(公路水平線形之一種)
Spiral Bar	螺旋鋼筋(簡稱Spiral)
Spiral Reinforcement	螺旋筋(用於圓混凝土柱)
Splice	搭接部份(例如兩個鋼樑之組接)
Splice Plate	鋼結構之接合鈑
Spliting Failure	劈裂破壞(對混凝土結構而言)

見下頁續…

英　　文	中　　文
Spread Footing	擴展式基腳
Spread Foundation	擴展式基礎
Spring Hammer Method	彈簧鎚法(非破壞檢測之一種)
Spring Line	拱橋之拱軸線
Spurs	丁壩
Square Crossing	直角式交叉，垂直交叉
Square Foundation	方型基礎
St. Venant Torsion	聖維南扭曲(與純扭曲，Pure Torsion同義)
Stability	穩定，穩定性
Stability Factor	穩定因數
Stability Number	穩定數(設計擋土牆時用之)
Stability Tower	穩定架(橋樑施工時之臨時穩定措施，亦有人以Stability Pier稱之)
Staged Construction	分段施工
Staggering	交錯排列
Staging Bent	臨時施工支架
Stainless Steel	不銹鋼
Standard Design	標準設計
Standard Deviation	標準差
Standard Penetration Resistance	標準貫入阻力

見下頁續…

英　　文	中　　文
Standard Penetration Test (SPT)	標準貫入試驗
Standard Sand	標準砂
Standard Specifications	標準規範
Star Cracks	星狀裂縫
Starling	橋墩臨上游部份擋水流與冰流之部份
Static Ice Pressure	靜冰壓力
Static Load	靜力
Statically Determinate Structure	靜定結構
Statically Indeterminate Structure	靜不定結構，超靜定結構
Station	公路里程之樁號(工程上以Sta.簡稱)
Steady-State Vibration	穩態振動
Steam Curing	蒸氣養護
Steel Angles	角鋼
Steel Bar	鋼棒，鋼筋
Steel Base Plate	鋼底鈑
Steel Box Girder	箱型鋼樑
Steel Bridge	鋼橋
Steel Cable	鋼索
Steel Casing	鋼製套管

見下頁續…

英　　文	中　　文
Steel Caisson	鋼製沈箱
Steel Conduits	鋼導管
Steel Fiber	鋼纖維
Steel Form	鋼模
Steel Girder Bridge	鋼梁橋
Steel Grid	鋼格床，鋼格梁
Steel Grid Plate	鋼格床鈑
Steel Piles	鋼樁
Steel Pipe Pile	鋼管樁
Steel Plate	鋼鈑
Steel Plate Bonding Method	鋼鈑黏貼法(混凝土構件補強方式之一)
Steel Reinforced Concrete (SRC)	鋼骨鋼筋混凝土
Steel Relaxation	鋼材鬆弛
Steel Ring	鋼環
Steel Sheet Piles	鋼鈑樁(簡稱Sheet Piles)
Steel Sliding Plate	鋼滑鈑
Steel Strand	鋼索
Steel Structures Painting Council (SSPC)	美國鋼結構油漆協會
Steel Tendon	鋼腱
Steel Top Plate	鋼頂鈑
Steel Wire	鋼線

見下頁續…

英　　　文	中　　　文
Stem	直壁(橋台、橋墩之主壁體)
Stiffener	加勁鈑，加勁材
Stiffening Rib	加勁肋
Stiffness	勁度
Stiffness Coefficient	勁度係數
Stiffness Constant	勁度常數
Stiffness Parameter	勁度參數
Stiffness Decay	勁度衰減
Stiffness Ratio	勁度比
Stirrup	箍筋
Stone Bridge	圬石橋(與Masonry Bridge同義)
Stopping Sight Distance	停車視距
Storm Cable	耐風索(與Wind Cable同義)
Straight Bridge	直線形橋樑
Straightening	直化法(鋼材維修方法之一)
Straight-Wingwall Abutment (Straight Abutment)	直翼型橋台(橋台主體與翼牆位於同一平面)
Strain Energy	應變能
Strain Gauge	應變計
Strain Gradient	應變梯度，應變升降率

見下頁續…

英　　文	中　　文
Strain Hardening Ratio	應變硬化比
Strands	鋼鉸索，鋼索(由數條鋼線扭轉而成，最常用者為7線鋼索，強度270ksi，參照ASTM A416)
Stream Flow Force Constant	水流力常數
Stream Flow Pressure	水流壓力
Strength Design Method	強度設計法(AASHTO橋梁設計方法之一，與Load Factor Design同義)
Strength Evaluation	強度評估
Strength Weld	耐力熔接，耐力焊接(與Stress Weld同義)
Strengthening	加固
Strength Reduction Factor	強度折減因數
Stress Concentration	應力集中
Stress Corrosion Crack	應力銹蝕裂縫
Stress Cycle	應力循環(用於疲勞分析)
Stress Distribution	應力分佈
Stress Range	應力範圍(用於疲勞分析)

見下頁續…

英　　文	中　　文
Stress Reversal	應力反覆(結構構件上某一點受壓應力之後，因荷重變化而變成拉應力之現象，反之亦然)
Stress-Strain Curve	應力－應變曲線
Stress Weld	耐力熔接，耐力焊接(與Strength Weld同義)
Stringer	縱樑，小樑(通常比Girder小)
Strip Seal Joint	帶狀填縫
Structural Analysis	結構分析
Structural Carbon Steel	結構碳鋼
Structural Concrete	結構混凝土
Structural Condition	結構物目前所處之狀況
Structural Cracks	結構裂縫
Structural Design	結構設計
Structural Engineer	結構工程師
Structural Engineering	結構工程
Structural Idealization	結構理想化
Structural Integrity	結構整體性
Structural Joint	結構縫，構造縫
Structural Key	結構榫(此種榫具有結構功能，例如增強剪力抵抗力)
Structural Members	具結構功能之構件

見下頁續…

英　　文	中　　文
Structural Modeling	結構模式
Structural Redundancy	結構贅餘度
Structural Responses	結構反應(例如彎矩、剪力等等)
Structural Specifications	結構規範
Structural Stability	結構穩定度
Structural Stability Research Council (SSRC)	美國之結構穩定研究委員會
Structural Steel	結構鋼
Structure Dimensions	結構尺寸
Stub Abutment	短塊式橋台
Stud Welding	植釘焊接
Submerged Arc Welding (SAW)	潛弧焊接
Substructure	下部結構
Sufficiency Rating	使用性評定
Super High Early Strength Portland Cement	超早強波特蘭水泥
Superelevation	公路橫斷面之超高
Superelevation Runoff	超高漸變段
Superimposed Dead Load	附加靜載重
Superplasticizer	強塑劑

見下頁續…

英　　　　文	中　　　　文
Superstructure	上部結構
Support Beam	支撐樑
Surcharge	超壓荷重
Surcharge Pressure	加載壓力
Surface Absorption	表面吸附
Surface Area	表面積
Surface Coating	表面塗裝
Sureface Corrosion	表面腐蝕
Surface Hardness Method	表面硬度法(非破壞檢測之一種)
Surface Layer	表面層，表層
Surface Roughness	表面粗糙度(構材處理完畢後之表面凹凸程度)
Surface Shear	表面剪力
Surface Strength Method	表面強度法(非破壞檢測之一種)
Surface Voids	表面孔隙
Suspension Bridge	懸索橋，吊橋
Swing Bridge	水平旋轉式橋樑(可動式橋樑其中一種)
Synthetic Rubber	合成橡膠
System Analysis	系統分析
System Identification	系統鑑定
Systematic Error	系統誤差

MEMORANDOM

英　　文	中　　文
T-Beam	T型樑
T-Girder	T型樑
T-Girder Bridge	T型樑橋
Tack Weld	點焊(通常指臨時性焊接以達到結構構件間之接續)
Temperature Adjustment	溫度調節
Temperature Crack	溫度裂縫
Temperature Gradient	溫度梯度
Temperature Reinforcement	溫度鋼筋
Temperature Strain	溫度應變
Temperature Stress	溫度應力(與Thermal Stress同義)
Temporary Bridge	臨時性橋樑
Temporary Connection	臨時接頭
Temporary Loads	臨時性荷重
Temporary Pier	臨時橋墩
Temporary Posttensioning Blister	臨時後拉預力凸塊
Temporary Support	臨時支撐
Temporary Strengthening	臨時性加固
Template	型版
Tendon	鋼腱
Tendon Curvature	鋼腱曲率

見下頁續…

英　　文	中　　文
Tendon Jacking Force	鋼腱加載力
Tendon Layout	鋼腱佈置
Tendon Pop-Out	鋼腱爆開
Tendon Relaxation	鋼腱鬆弛
Tendon-Truss Element	鋼腱－桁架元素
Tensile Strength	拉力強度，抗拉強度
Tensile Strength Test	拉力強度試驗(非破壞檢測方法之一種)
Tensile Stress	拉應力，張應力
Tensile Stress Path	拉應力路徑
Tensile Zone	受拉區
Tension	拉力，張力
Tension Crack	拉力裂縫
Tension Flanges	受拉翼版
Tension Field Action	拉力場作用
Tension Member	受拉構件
Tension Tendon	拉力鋼腱
Testing Pistol Method	測試槍法(非破壞性檢測之一種)
Test Piles	試樁
Thermal Differential	溫差
Thermal Rise and Fall (TRF)	溫度升降(在設計節塊式混凝土橋樑時甚為重要)
Thermal Setting	熱凝
Thermometer	溫度計(橋樑檢測用之)

見下頁續…

英　　文	中　　文
Thin-Walled Beam	薄壁樑
Thread Bar	螺紋鋼棒
Three-Hinged Arch	三鉸拱
Three Point Support	三點式支撐(預鑄混凝土節塊儲存方式之一種)
Throat Cracks	焊喉裂縫(針對鋼結構而言)
Tidal Zone	潮汐區
Tie	拉桿，拉條，繫條
Tie Beam	連繫樑(用於連結相鄰之單腳式基腳)
Tie Rods	拉桿
Tilt	傾斜
Tiltmeter	傾斜量測計(橋樑檢測用之)
Timber Bridge	木橋
Timber Deck	木橋版
Timber Piles	木樁
Time Dependent Behavior	時變行為
Time Domain	時間域
Tire Contact Area	輪胎接觸面
Toe	趾版(擋土牆基礎之前端)
Toe of Slope	坡趾(邊坡之趾部)
Tolerance	可容許之尺寸誤差
Toll Road	收費道路

見下頁續…

英　　文	中　　文
Top Chord	頂弦桿(針對桁架而言)
Top Cover Plate	上蓋鈑
Top Flange	頂版，上翼版
Top Lateral Bracing	上側支撐(用於箱型鋼樑)
Top View	頂視圖
Torque	扭曲(與Torsion及Twisting同義)
Torque Coefficient	扭曲係數
Torsion Constant	扭曲常數
Torsion Crack	扭曲裂縫
Torsion Reinforcement	扭曲鋼筋，抗扭鋼筋
Torsional Cracking Moment	扭曲裂縫彎矩
Torsional Divergence	扭曲發散(大跨度橋樑受風力影響而產生之結構行為之一)
Torsional Displacement	扭曲變形
Torsional Shear Stresses	扭曲剪應力
Torsional Shear Test	扭剪試驗
Torsional Stiffness	扭曲勁度
Total Normal Stress	總正向應力
Total Prestress Loss	總預力損失
Total Strain Energy	總應變能
Toughness	韌性
Toughness Index	韌性指數

見下頁續…

英　　文	中　　文
Tower-Type Pier	塔式橋墩(工程界以Tower簡稱之)
Traffic Composition	交通組成
Traffic Control	交通控制
Traffic Control Plan	交通控制計畫
Traffic Flow	車流
Traffic Impact Accessment (TIA)	交通衝擊評估
Traffic Lanes	車道
Traffic Lighting	交通號誌
Traffic Loads	交通荷重
Traffic Marking	標線(繪於公路路面者)
Traffic Volume	交通量
Train Loading	列車載重
Transition Curve	介曲線，緩和曲線(公路平面線形之一種)
Transportation Research Board (TRB)	美國之運輸研究委員會(隸屬於National Research Council之下)
Transverse Analysis	橫向分析
Transverse Beam	橫樑
Transverse Bending	橫向彎矩
Transverse Bending Stiffness	橫向彎矩勁度
Transverse Bracin	橫向支撐

見下頁續…

英　　文	中　　文
Transverse Cracks	橫向裂縫
Transverse Earthquake Motion	橫向地震
Transverse Expansion Joints	橫向伸縮縫
Transverse Force	橫向力
Transverse Members	橫向構件
Transverse Posttensioning	橫向後拉預力
Transverse Reinforcement	橫向鋼筋
Transverse Ribs	橫向肋樑
Transverse Stiffener	橫向加勁版，橫向加勁材
Transverse Tensile Stress	橫向拉應力
Transverse Torsion	橫向扭曲，橫向扭矩
Treated Timber	處理過之木材
Tremie Concrete	特密混凝土
Tremie Pipe	特密管
Trestle Bridge	棧橋
Tiral and Error Method	試誤法
Triaxial Compression Test	三軸壓縮試驗
Truck Load Test	卡車載重試驗(橋樑強度評定方法之一)

見下頁續…

英　　文	中　　文
Truck Loading	貨車載重(設計橋樑時之活載重之一)
Trumpet	錐形管(用於預力混凝土結構)
Truss	桁架
Truss Bridge	桁架橋
Tubular Piers	管狀橋墩
Tuned Mass Damper (TMD)	調質阻尼器
Tunnel	隧道
Turbulence	亂流(指風而言)
Turbulence Intensity	亂流強度
Turnbuckle	相鄰鋼腱或鋼棒之接頭
Twin-Column Pier	雙柱式橋墩
Two-Stage Casting	兩段式澆注
Two-Way Posttensioning	雙向後拉預力
Two-Way Punching Shear	雙向剪穿力
Two-Way Slab	雙向版

MEMORANDOM

英　　文	中　　文
U-Bolt	U型螺栓
U-Type Abutment	U型橋台
Ultimate Axial Load Capacity	極限軸向荷重強度
Ultimate Bearing Capacity	極限承載力
Ultimate Drying Shrinkage	極限乾縮
Ultimate Load	極限載重，極限荷重
Ultimate Load Capacity	極限載重容量
Ultimate Pile Capacity	極限樁強度
Ultimate Resistability	極限抗力
Ultimate Shrinkage Crack	極限乾縮裂縫
Ultimate Shrinkage Strain	極限乾縮應變
Ultimate Strength	極限強度
Ultimate Strength Design (USD)	極限強度設計(與LFD同義)
Ultimate Tensile Strength	極限抗拉強度
Ultrasonic Pulse	超音波脈波
Ultrasonic Pulse Velocity Method	超音波脈波速度法(非破壞檢測中，音波檢測法之一種)

見下頁續…

英　　文	中　　文
Ultrasonic Testing (UT)	超音波試驗，超音波檢測
Unbalanced Moment	不平衡彎矩
Unbonded Length	非黏結長度(地錨系統前段，其後段為Bonded Length)
Unbonded Tendon	非黏結性鋼腱
Unbraced Length	未支撐長度
Unbraced Section	側向支撐不足斷面(AASHTO之LFD鋼橋設計中之某一種斷面型式)
Unconfined Compression Strength	單軸壓縮強度
Unconfined Compression Test	單軸壓縮試驗
Unconfined Flow	無側限流
Underconsolidated Clay	壓密中黏土
Undercut	勾邊(鋼結構焊接之缺陷之一)
Underdrain	排水用之暗管、盲管
Undermind Foundation	被淘空之基礎
Underflow	潛流
Undermining	淘空
Underpass	穿越
Underpass Bridge	穿越橋

見下頁續…

英　　文	中　　文
Underpinning	水中補砌工法(橋樑下部結構維修方法之一)
Underreinforced Concrete	低筋混凝土
Underwater Concreting	水中澆灌混凝土
Underwater Inspection	水中檢測
Underwater Photography	水中照相(於水中檢測時，為紀錄橋樑下部結構之狀況而進行照相)
Unfilled Tubular Steel Piles	中空鋼管樁
Uniform Load	均佈載重，均佈荷重
Uniform Torsion	均勻扭曲
Uniformity Coefficient (UC)	均勻係數
Unit-Load Method	單位力法
Unsymmetrical Section	不對稱斷面
Uplift Bearing	抗拉支承
Uplift Forces	上舉力，上拉力
Upper Bound	上限
Upper Bound Envelopes	上限包絡線
U.S. Army Corps of Engineer	美國工兵團
US DOT	美國聯邦之運輸部

英　　文	中　　文
Value Engineering (VE)	價值工程
Variable Depth Boxgirder	梁深漸變之箱型梁
Vaulted Abutment	地窖式橋台(與Cellular Abutment同義)
Vehicle-Bridge Interaction	車輛橋梁互制
Velocity Censors	速度規
Vent	通氣口
Vertical	垂直縱向鋼筋(尤指RC柱)之簡稱
Vertical Alignment	縱斷線形
Vertical Clearance	淨高，淨空
Vertical Cracks	垂直裂縫
Vertical Curvature	縱斷曲率
Vertical Curve	豎曲線(公路幾何線形之一種)
Vertical Displacement	垂直變形
Vertical Ground Motion	垂直地表運動
Vertical Lift Bridge	垂直升降式橋梁
Vertical Posttensioning	垂直後拉預力
Vertical Reaction	垂直反力
Vertical Reinforcement	垂直鋼筋
Vertical Seismic Coefficient	垂直地震力係數

見下頁續…

英　　文	中　　文
Vertical Shear	垂直剪力
Vertical Slip Form	垂直滑動模板
Vertical Strain	垂直應變
V-Heating	V型加熱法(利用加熱使鋼樑彎曲之方法之一)
Viaduct	高架路
Vibrating Hammer	震動錘
Vibration	振動
Vibrator	振動器
Virtual Work	虛功
Viscous Damper	黏滯性阻尼器
Void	空隙，孔隙
Void Ratio	空隙比，孔隙比
Voided Slab	中空版
Voided Slab Bridge	中空版橋
Volumetric Ratio	體積比
Vortex-Induced Oscillation	渦激振動(用於橋樑之風力分析)

英　　文	中　　文
Wading Inspection	涉水徒步檢測
Wall Footing Foundation	牆式基腳基礎
Wall-Type Pier	牆式橋墩
Warning Sign	警告標誌
Warping	翹曲(薄壁結構受扭曲荷重之結構反應之一種)
Warping Constant	翹曲常數
Warping Deformation	翹曲變形
Warping Effect	翹曲效應
Warping Function	翹曲函數
Warping Moment	翹曲彎矩(與雙彎曲，Bimoment，同義)
Warping Normal Stresses	翹曲正向應力
Warping Rigidity	翹曲剛度
Warping Shear	翹曲剪力
Warping Shear Decay Coefficient	翹曲剪力衰減係數
Warping Torsion	翹曲扭曲(與不均勻扭曲同義)
Washer	墊圈(用於鋼結構之螺栓中)
Water Buoyancy Factor	水浮力因數
Water Cement Ratio (WCR)	水灰比

見下頁續…

英　　　文	中　　　文
Water-Jet Blasting	高壓水噴砂(清除鋼鐵表面異物、雜質之方法之一)
Water Level	水位
Water Pressure	水壓力
Water Surface	水面
Waterway	水道，水路
Waterway Bridge	水路橋，輸水橋
Waterproofing Material	防水材料
Waterproofing Membrane	防水膜
Water-Reducing Agent	減水劑
Water-Reducing Retarder	減水緩凝劑
Waterstop	止水帶(用於結構物之接縫處)
Watertightness	水密性
Waterway	水道
Wave Equation	波動方程式
Weak Axis	弱軸
Wearing Surface	磨耗層(舖裝層)
Wearing Surface Treatment	磨耗層處理
Weathering	風化
Weathering Cracks	天候裂縫
Weathering Steel	抗蝕鋼，防蝕鋼，耐蝕鋼(例如A588鋼材)

見下頁續…

英　　文	中　　文
Web (Web Plate)	腹版
Web Buckling Coefficient	腹版挫屈係數
Web Key	腹版剪力榫(通常位於預鑄混凝土箱型樑之腹版)
Web Opening	腹版開孔
Web Reinforcement	腹筋
Web Stiffener	腹版加勁材
Wedge	鋼楔，楔子(用於預力混凝土結構端錨處)
Wedge Plate	楔版
Weeper	排水洞
Weep Hole	洩水孔、排水孔
Weld Crack	焊接裂縫
Weld Sequence	焊序
Weldable Steel	可焊接鋼
Welded Wire Mesh	焊接鋼絲網
Welding	焊接
Welding Electrode	焊條(常以Electrode簡稱之)
Welding Pass	焊道(對焊接而言)
Wet Construction Method	濕式施工法
Wet Corrosion	濕腐蝕(亦稱電化腐蝕)
Wet Sand Blasting	濕式噴砂(清除鋼鐵表面異物、雜質之方法之一)

見下頁續…

英　　　文	中　　　文
Wetland	濕地(橋樑設計時必須特別注意與環保單位聯繫討論之)
Wheel Base	車輛之軸距
Wheel Load	輪重(對直線形橋樑而言，輪重為軸重之半)
Wheel Load Distribution Constant	輪荷重傳遞常數
Wind Bracing	風拉桿
Wind Cable	耐風索(與Storm Cable同義)
Wind Load	風荷重
Wind Resistant Design	抗風設計
Wind Tunnel	風洞
Wind Uplift	風上揚力
Wingwall	翼牆(亦有人以Wing簡稱之)
Wingwall-Type Abutment	翼牆型橋台
Winter Concrete	冬天混凝土
Wire Brusher	鐵絲刷(用於除銹)
Wires	鋼線(通常指用於預力混凝土結構之鋼絲線，常用之鋼線直徑約 1/4英吋左右，參照ASTM A421)
Work Platform	工作台，工作平台

見下頁續…

英　　文	中　　文
Workability	工作度
Working Load	工作荷重(與服務載重，Service Load，同義)
Working Stress Design (WSD)	工作應力設計(AASHTO中設計方法之一種，與ASD同義)

英　　文	中　　文
Yield	公路上"讓行"之標誌
Yield Line Analysis	降伏線分析
Yield Point	降伏點
Yield Strength	降伏強度
Yield Stress	降伏應力
Yielding Walls	柔性牆(可以牆底為基準而前傾或旋轉之擋土牆)

國家圖書館出版品預行編目資料

橋樑工程英漢辭典 / 徐耀賜著. -- 初版. --
臺北市：全華，民88
面；　公分
ISBN 957-21-2556-7 (平裝)

1. 橋樑工程 - 字典，辭典　2. 英國語言 -
字典，辭典 - 中國語言

441.804　　　　88005573

橋樑工程英漢辭典

徐　耀　賜　著
執行編輯／田　惠　敏
封面設計／周　雍　勝
發 行 人／詹　儀　正
出 版 者／全華科技圖書股份有限公司
地址：台北市龍江路76巷20-2號2樓
電話：25071300（總機）FAX：25062993
郵撥帳號：0100836-1號
印 刷 者／宏懋打字印刷股份有限公司
登 記 證／局版北市業字第○七○一號
圖書編號／03617
ＩＳＢＮ／957-21-2556-7
定　　價／新臺幣 200 元
初版一刷／88年6月

全華網際中心 URL　http://www.chwa.com.tw
E-mail　book@msl.chwa.com.tw

全華科技圖書 讀者服務卡

為加強對您的服務,只要您填妥本卡寄回本公司(免貼郵票)即可成為全華讀友並長期享受新書介紹及各種促銷活動!

填寫日期: 年 月 日

姓　　名		生　　日	年　月　日
E-mail			
需求書類	□A 電子 □B 電機 □C 計算機工程 □D 資訊 □E 機械 □F 汽車 □I 工管 □J 土木 □K 化工 □L 美工		
教育程度	□國中 □高中 □專科 □大學 □研究所		
職　　業	□工程師 □教師 □學生 □軍 □公 □其他 學校/公司:______ 科系/部門:______		
電　　話	公司:	住家:	傳真:
地　　址	□□□		
購買圖書	書號:	書名:	
購買動機	□逛書局 □雜誌廣告 □報紙廣告 □老師推薦 □書展 □同學介紹 □其他______		
購買方式	□總公司門市 □全友書局 □團體購買 □郵購 □書展 □網路訂購 □其他______		
內容評價			
建　　議	希望全華出版: 希望全華加強那些服務:		

※請詳填,並書寫端正,謝謝!

88.03.300,000

親愛的讀者:

感謝您對全華圖書的支持與愛用,雖然我們很慎重的處理每一本書,但疏漏之處在所難免,若您發現本書有何錯誤之處,請填寫於勘誤表內,我們將於再版時修正。您的批評與指教是我們前進的最大動力,謝謝您!

全華圖書敬上

勘誤表

書　號		書　名		作　者	
頁　數	行　數	錯誤或不當之詞句	建議修改之詞句		

我有話要說:(其他之批評與建議,如:封面、編排、內容、印刷品質.... 等)

加入全華書友就是那麼容易

1. 親至本公司購書三本以上者
   請直接向門市人員提出申請。
2. 劃撥購書一次滿三本以上者
   請在劃撥單通訊欄註明申請書友證。
3. 填妥讀者服務卡三張寄回本公司(免貼郵票)。

只要符合以上條件之一，即可獲得全華書友證乙張
同時在您提出申請的二週內即可收到書友證。

成為全華書友的好處：

1. 長期享受中文新書 8 折優惠(進口西書 95 折)。
2. 定期獲贈全華最新出版與專案促銷活動訊息。
3. 享有生日禮讚特惠驚喜活動。

全華專業科技圖書網站：http://www.chwa.com.tw
E-mail: service@ms1.chwa.com.tw

廣　告　回　信
台灣北區郵政管理局登記證
北台字第 1689 號

104
台北市中山區龍江路76巷20之2號2樓
全華科技圖書股份有限公司
行銷企劃部 收